Mastering Blazor WebAssembly

A step-by-step guide to developing advanced single-page
applications with Blazor WebAssembly

Ahmad Mozaffar

BIRMINGHAM—MUMBAI

Mastering Blazor WebAssembly

Copyright © 2023 Packt Publishing

Group Product Manager: Rohit Rajkumar

Publishing Product Manager: Jane D'Souza

Senior Editor: Divya Anne Selvaraj

Technical Editor: Saurabh Kadave, Joseph Aloocaran

Copy Editor: Safis Editing

Language Support Editor: Safis Editing

Project Coordinator: Sonam Pandey

Proofreader: Safis Editing

Indexer: Subalakshmi Govindhan

Production Designer: Vijay Kamble

Marketing Coordinator: Nivedita Pandey

First published: August 2023

Production reference: 1180723

Published by Packt Publishing Ltd.

Grosvenor House

11 St Paul's Square

Birmingham

B3 1R

ISBN 978-1-80323-510-3

www.packtpub.com

Huge thanks to all my awesome family: my father, Khaled, and my mother, Samar, for raising me with a love of science, and teaching me the power of learning and giving. To my brothers, Dr. Waddah and Rasheed, for being my best friends and a source of inspiration throughout my life since day one.

– Ahmad Mozaffar

Contributors

About the author

Ahmad Mozaffar is a senior software engineer, cloud developer, and trainer who currently works at ContraForce as a senior full stack engineer responsible for developing SaaS platforms using Blazor, Azure, AI, and other modern technologies. Originally from Syria, his passion for software development, especially .NET and Azure, started at a very early age. He loves teaching people about developing software using .NET, Azure, and other Microsoft technologies through his YouTube channel, AK Academy, and writing articles on multiple platforms, which have reached more than 400K readers.

About the reviewer

Steven Giesel is a seasoned .NET developer and Microsoft MVP in the "Developer Technologies" category with over nine years of professional experience. He is also the (co-)maintainer of multiple open source libraries, among them the famous bUnit library, which is used for Blazor unit testing and is discussed in this very book.

Table of Contents

3

Developing Advanced Components in Blazor   59

Part 2: App Parts and Features

4

Navigation and Routing   91

5

Capturing User Input with Forms and Validation 111

6

Consuming JavaScript in Blazor 133

10

Handling Errors in Blazor WebAssembly   235

Part 3: Optimization and Deployment

11

Giving Your App a Speed Boost   253

12

RenderTree in Blazor   269

13

14

15

What's Next? 325

Index 335

Other Books You May Enjoy 346

Preface

Single-page applications (**SPAs**) have become an essential part of modern technology. We have become more dependent on browsers than ever before. With the release of Angular, ReactJS, and other JavaScript frameworks for developing modern web apps, a big revolution occurred. JavaScript was dominant in the field for many years as the only language to be understood and run in browsers. In 2015, WebAssembly arrived as a technology experiment that allowed browsers to run programming languages other than JavaScript, such as C# and C.

With WebAssembly, the doors have opened for more powerful web apps and new frameworks to be used to build things in browsers. Microsoft was the first adopter of WebAssembly by announcing Blazor, a framework that allows .NET developers to build modern and advanced web applications using C# that can run natively in a browser.

In 2018, when Blazor was still an experimental project of Microsoft's, I started to use it, and I was amazed at how powerful it was and the huge potential it had in the world of web and software development in general, outside the boundaries of the browser. From that day onward, I started to create packages and build demos and production-ready apps with it, until it became the main technology I used for my full-time work.

Mastering Blazor WebAssembly utilizes all my experience in Blazor and .NET to show you, in depth, how Blazor WebAssembly works and how you can leverage it, step by step, to build fascinating modern web apps that run natively with C# inside the browser. It will show you how to build the foundations you need to use Blazor WebAssembly to build cross-platform mobile and desktop apps by utilizing .NET MAUI and Blazor WebAssembly.

The book begins by introducing Blazor WebAssembly and how to set up a new project. Then, it covers all the features and concepts of Blazor WebAssembly with practical examples and theoretical explanations. You will learn how Blazor WebAssembly works internally and how to use it effectively. By the end of this book, you will have developed a complete, workable, and efficient test application that you can publish to the cloud.

Who this book is for

This book targets existing .NET developers who are excited to start their journey with developing SPAs, using their own C# skills without learning JS frameworks, JS developers who have heard about Blazor and want to discover its superpower and simplicity for web development, and every geek who wants to discover and learn how things work under the hood, step by step.

What this book covers

Chapter 1, Understanding the Anatomy of a Blazor WebAssembly Project, introduces Blazor WebAssembly, guides you to set up a project and understand its structure, and covers essential topics such as environments and dependency injection.

Chapter 2, Components in Blazor, explains the concept of components in Blazor WebAssembly and SPAs, covering how to build, communicate, and style them.

Chapter 3, Developing Advanced Components in Blazor, introduces advanced component types such as layouts, templated components, and dynamic components, showing you how to use Razor class library projects to build reusable components.

Chapter 4, Navigation and Routing, explains the navigation process in a SPA and how to implement it in your Blazor apps. The chapter also covers how to use query parameters, send and receive data via a URL, and respond to navigation changes.

Chapter 5, Capturing User Input with Forms and Validations, explains how to create and submit forms in Blazor, use the built-in input components, validate user input, and create your own custom input component.

Chapter 6, Consuming JavaScript in Blazor, explains how and when to use JS in your Blazor WebAssembly project and how to call JS code from your Blazor app, and vice versa. It also shows how to wrap an existing JS package into a Blazor component using JS.

Chapter 7, Managing Application State, introduces the concept of state management in an SPA and shows three different techniques for preserving the state of your application – using local storage, in memory, or a URL.

Chapter 8, Consuming Web APIs from Blazor WebAssembly, provides an overview of web APIs and web API clients, showing you how to understand an existing web API and send HTTP requests to it from Blazor. It also covers delegating handlers, `IHttpClientFactory`, and how to organize API calls in your app.

Chapter 9, Authenticating and Authorizing Users in Blazor, explains what authentication and authorization are, and then deep-dives into how Blazor handles authentication and how to develop your custom authentication flow with JSON Web Tokens. It also shows you how to control the UI and the app logic based on the authentication state and call secured web API endpoints.

Chapter 10, Handling Errors in Blazor WebAssembly, shows you how to make your app reliable and the techniques to handle errors efficiently in your apps.

Chapter 11, Giving Your App a Speed Boost, provides you with advanced mechanisms in Blazor that help you make your app as efficient as possible, by increasing the rendering speed and reducing the app size for faster load time.

Chapter 12, RenderTree in Blazor, explains in detail how Blazor renders components and manages the **Document Object Model** (**DOM**), introducing the concept of `RenderTree`. It also shows how you can improve your app performance by learning how things work under the hood.

Chapter 13, Testing Blazor WebAssembly Apps, introduces the concept of component testing and the bUnit library, providing you with what you need to write efficient unit tests for your components. The chapter also covers **End-to-End** (**E2E**) testing and the Playwright package.

Chapter 14, Publishing Blazor WebAssembly Apps, shows the checks to do before publishing an app, introduces the Blazor WebAssembly ASP.NET Core Hosted app model, and guides you step by step to release your apps to Azure App Service and Azure Static Web Apps.

Chapter 15, What's Next?, provides an overview of more components and use cases to help you apply what you learned in this book.

To get the most out of this book

You need to have a basic understanding of HTML and CSS, in addition to being familiar with *.NET* and *C#* and the concept of *Object-Oriented Programming* (*OOP*). No prior experience in *JS* is needed.

Software/hardware covered in the book	Operating system requirements
Visual Studio 2022	Windows
Visual Studio Code	Windows, macOS, or Linux
.NET 7.0	Windows, macOS, or Linux

In *Chapter 14, Publishing Blazor WebAssembly Apps*, an Azure account is needed to follow along with the chapter.

If you are using the digital version of this book, we advise you to type the code yourself or access the code from the book's GitHub repository (a link is available in the next section). Doing so will help you avoid any potential errors related to the copying and pasting of code.

Download the example code files

You can download the example code files for this book from GitHub at `https://github.com/ PacktPublishing/mastering-blazor-webassembly`. If there's an update to the code, it will be updated in the GitHub repository.

We also have other code bundles from our rich catalog of books and videos available at `https:// github.com/PacktPublishing/`. Check them out!

Conventions used

There are a number of text conventions used throughout this book.

`Code in text`: Indicates code words in text, database table names, folder names, filenames, file extensions, pathnames, dummy URLs, user input, and Twitter handles. Here is an example: "Then, the user is authenticated if the `user` object inside the `UserState` class is not `null`."

A block of code is set as follows:

```
<nav class="navbar navbar-expand-lg navbar-light bg-light">
    ...
    <div class="d-flex">
        <button class="btn btn-outline-primary">
            Login</button>
    </div>
    </div>
</nav>
```

Any command-line input or output is written as follows:

```
dotnet build
dotnet run
```

Bold: Indicates a new term, an important word, or words that you see on screen. For instance, words in menus or dialog boxes appear in **bold**. Here is an example: "Another example we will go through is the **Add to Cart** button."

> **Tips or important notes**
> Appear like this.

Get in touch

Feedback from our readers is always welcome.

General feedback: If you have questions about any aspect of this book, email us at customercare@ packtpub.com and mention the book title in the subject of your message.

Errata: Although we have taken every care to ensure the accuracy of our content, mistakes do happen. If you have found a mistake in this book, we would be grateful if you would report this to us. Please visit www.packtpub.com/support/errata and fill in the form.

Piracy: If you come across any illegal copies of our works in any form on the internet, we would be grateful if you would provide us with the location address or website name. Please contact us at copyright@packt.com with a link to the material.

If you are interested in becoming an author: If there is a topic that you have expertise in and you are interested in either writing or contributing to a book, please visit authors.packtpub.com.

Share Your Thoughts

Once you've read *Mastering Blazor WebAssembly*, we'd love to hear your thoughts! Scan the QR code below to go straight to the Amazon review page for this book and share your feedback.

https://packt.link/r/1-803-23510-1

Your review is important to us and the tech community and will help us make sure we're delivering excellent quality content.

Download a free PDF copy of this book

Thanks for purchasing this book!

Do you like to read on the go but are unable to carry your print books everywhere?

Is your eBook purchase not compatible with the device of your choice?

Don't worry, now with every Packt book you get a DRM-free PDF version of that book at no cost.

Read anywhere, any place, on any device. Search, copy, and paste code from your favorite technical books directly into your application.

The perks don't stop there, you can get exclusive access to discounts, newsletters, and great free content in your inbox daily

Follow these simple steps to get the benefits:

1. Scan the QR code or visit the link below

https://packt.link/free-ebook/9781803235103

2. Submit your proof of purchase
3. That's it! We'll send your free PDF and other benefits to your email directly

Part 1: Blazor WebAssembly Essentials

This part will introduce you to Blazor WebAssembly and guide you to set up your first Blazor project. After that, you will learn about the concept of components, which are the building blocks of every Blazor app. You will start by using Razor and C# to create basic and advanced components, and then you will learn how to use parameters in Blazor to implement communication between app components, and how to style them. You will also learn about different types of components, such as layouts and templated components, and how to employ a Razor class library for reusability.

This part has the following chapters:

1

Understanding the Anatomy of a Blazor WebAssembly Project

Blazor has become a trend in the world of software development, due to its simplicity, productivity, and wide-ranging applicability in web, mobile, and desktop application development via **.NET Multi-platform App UI (MAUI)** (more about MAUI in *Chapter 15, What's Next?*). Thus, it is beneficial for developers to learn about and master this framework.

This book will help you master Blazor WebAssembly by building a new project in Visual Studio until you have a fully functional app that is secure, robust, deployed online, and ready to be consumed by the public.

In this chapter, we are going to walk you through the basic structure of a Blazor WebAssembly standalone application. We will start by creating the project using the various methods available on different platforms, including the *Visual Studio IDE, Visual Studio Code*, and the *.NET CLI*. You will then learn about the general structure of the project and what each file contains and why it's there. You will also learn about **dependency injection (DI)**, how to register and inject services, and how to store and retrieve the configuration of your application in single or multiple environments.

By the end of this chapter, you will have a good idea of the structure of the application we will be building in the later chapters of this book.

In this chapter, we will cover the following topics:

- Creating your first Blazor WebAssembly project
- Discovering the project structure
- Dependency injection in Blazor WebAssembly
- Creating, storing, and retrieving the app configuration
- Managing application environments

Technical requirements

To follow along with this chapter and the rest of the book, you need to know the basic concepts of the **C# programming language** and the fundamentals of **web** development.

The development process will require *Visual Studio 2022 Community*, which can be downloaded for free for *Windows* devices from `https://visualstudio.microsoft.com/vs/community`, and for *macOS* devices from `https://visualstudio.microsoft.com/vs/mac/`.

Alternatively, you can use *Visual Studio Code* for any OS (Windows, macOS, or Linux) by downloading it from `https://code.visualstudio.com/download`.

If you are not going to be installing *Visual Studio 2022*, you need to make sure that the *.NET 7.0 SDK* is available on your computer. This SDK generally gets downloaded along with Visual Studio 2022. If you don't have the SDK, you can install it from `https://dotnet.microsoft.com/en-us/download`.

In later chapters, we will mostly be using *Visual Studio 2022* for *Windows*, but all the concepts will be applicable to Visual Studio for Mac and *Visual Studio Code* with the *.NET CLI*.

You can find the complete source code for this chapter at `https://github.com/PacktPublishing/Mastering-Blazor-WebAssembly/tree/main/Chapter_01/BooksStore`.

Creating your first Blazor WebAssembly project

Let us start our journey by creating our Blazor WebAssembly project using the *.NET CLI*, and then we will do the same using *Visual Studio 2022* and will build on top of that for the rest of the book.

Using the .NET CLI

To create a Blazor WebAssembly project using .NET CLI, go through the following steps:

1. Open *Command Prompt* on *Windows* or *Terminal* on *macOS* and execute the following command:

    ```
    dotnet new blazorwasm -n BooksStore
    ```

    The preceding command will create a new Blazor WebAssembly project in a folder called `BooksStore` within the directory you executed this command in, and the project will be called `BooksStore`.

2. You can build your app using the `build` command after navigating to the newly created folder as follows:

    ```
    cd BooksStore\
    dotnet build
    ```

3. Next, you can run the project:

    ```
    dotnet run
    ```

When the app runs, you will see two URLs on the CLI: one for the *https* link and the other for an *http* link. Navigate to the https link in your browser and you will be able to see the final result of the default Blazor WebAssembly app made available out of the box by *.NET 7.0*:

```
info: Microsoft.Hosting.Lifetime[14]
      Now listening on: https://localhost:7166
info: Microsoft.Hosting.Lifetime[14]
      Now listening on: http://localhost:5166
info: Microsoft.Hosting.Lifetime[0]
      Application started. Press Ctrl+C to shut down.
info: Microsoft.Hosting.Lifetime[0]
      Hosting environment: Development
info: Microsoft.Hosting.Lifetime[0]
      Content root path: C:\AK\Mastering Blazor WebAssembly\Source Code\Mastering-Blazor-WebAssem
bly\Chapter_01\BooksStore\BooksStore
```

Figure 1.1 – Output after running the dotnet command in Command Prompt

Using Visual Studio 2022

Now, let's take a quick look at how to create the same Blazor WebAssembly project using Visual Studio 2022:

1. Open Visual Studio 2022 and, from the starter page, click on **Create a new project**. Then, search for the `Blazor WebAssembly` template.

2. Select **Blazor WebAssembly App**:

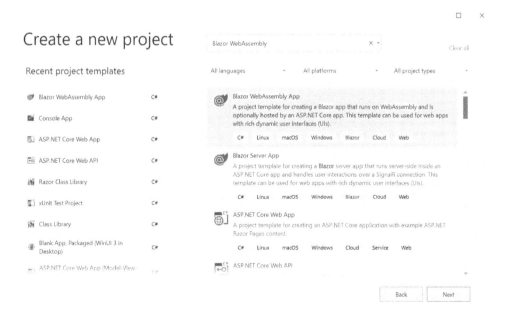

Figure 1.2 – Selecting the project template in VS 2022

3. Give your project a name and choose the directory you want to store it in on your machine:

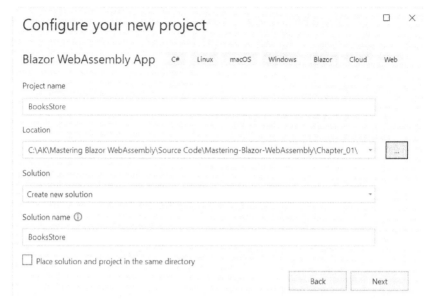

Figure 1.3 – Adding a project name and location

4. Keep the initial configuration as suggested by Visual Studio. Click on **Create**:

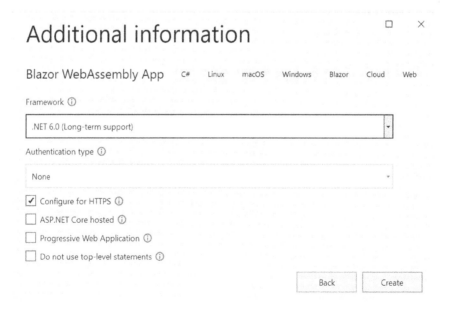

Figure 1.4 – Default project configuration in Visual Studio 2022

The project will be created, and you will see the following files and folders in your VS **Solution Explorer** panel:

Figure 1.5 – Project files and folders in the Solution Explorer panel

Clicking the *Start* (*F5*) button while in your VS 2022 window will run the project and open the browser automatically for you, and you will see the default Blazor app.

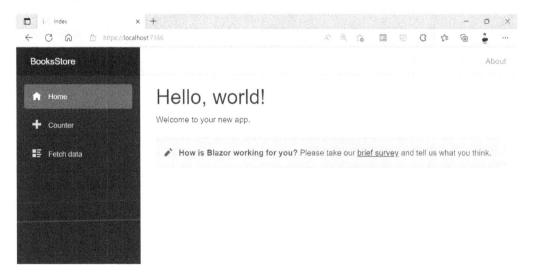

Figure 1.6 – Default running Blazor WebAssembly project

Congratulations! You have seen the famous purple UI; now, we are ready to move ahead and start discovering the purpose of each of those files and folders.

Discovering the project structure

Now we have a workspace that we can use to start building our application, but before we dive in deeper, let's take a look at the structure of the project, including the files and folders we see. Understanding the what and why of each file in **Solution Explorer** is key to completing the project we will be building.

Let's start exploring!

The wwwroot folder

The wwwroot folder is considered the root folder of your project. Root, here, doesn't mean the root in your file explorer, which is the one that contains the **Razor** files and other folders. wwwroot is the place where your app's static files, including global CSS styles, the main HTML file of the project, index.html, JavaScript, and assets (images, icons, static documents, etc.), live. It's also where you put the JSON files that represent the configuration of your app (you will read more about configuration in the next section of this chapter, *Dependency injection in Blazor WebAssembly*).

By default, the wwwroot folder contains the following:

- The css folder: This folder comes with Bootstrap **CSS** files and the default font. The folder also includes the app.css file, which carries the global styles of the app.

- The sample-data folder: This folder contains a **JSON** file with sample data pertaining to a weather forecast. This is used as a data source for the **Fetch data** page, which will be mentioned soon.

- favicon.ico: This refers to the default icon of the app that you see at the top left of the browser tab of your application, in the favorite sites list in the browser, and in some other places.

- Icon-192.png: This is the Blazor icon PNG file that's referenced in the components.

- index.html: The name of this file should be familiar to you. Most of the time, index.html is the default HTML file that will be opened when we access a static website. This main HTML file contains our app. The browser downloads this file and opens it with all its references from CSS and JavaScript, most importantly blazor.webassembly.js.

  The body of index.html contains, by default, a div with the ID app (this is basically where all the UI that you see in any Blazor app resides), and another div with the ID blazor-error-ui, which is a simple design div that shows up when an unhandled exception occurs (you will learn more about this in *Chapter 10, Handling Errors in Blazor WebAssembly*).

  While developing your apps, you are going to make changes to this file mostly to add style references or some JavaScript files later in this book.

The following is the code for the default body elements (app div and blazor-error-ui div) within the index.html file:

```
<html lang="en">
....
<body>
    <div id="app">Loading...</div>
    <div id="blazor-error-ui">
        An unhandled error has occurred.
        <a href="" class="reload">Reload</a>
        <a class="dismiss">✕</a>
    </div>
    <script
      src="_framework/blazor.webassembly.js"></script>
</body>

</html>
```

> **Note**
>
> The term **Single-Page Application (SPA)** seems a bit weird when you open the app and start navigating from one page to another. However, technically SPAs consist of a single HTML page and Blazor or any other JavaScript framework (ReactJS, Angular, etc.) dynamically updates the content of that page. As you have seen, in the wwwroot folder, there is only one HTML page in our whole application, called index.html.

The Pages folder

Pages is the default folder where Microsoft puts its default pages that come with the template. You can use Pages as the container for your app's page components. However, it's not mandatory to put your components in this folder and you can even delete it. Pages comes with three sample pages that cover different concepts, such as calling a shared component, updating the DOM using a button, and fetching data from a data source:

- Counter.razor: This is a Razor component that you can access through the route / counter. This is also an example of calling a C# method from an HTML button and binding an HTML <p> element's content to a C# variable to show the result.

- FetchData.razor: You can access this component by navigating to /fetch-data. This page contains an HTML table that shows weather forecast data and demonstrates how to fetch data using the HttpClient. By injecting the object and making an HTTP GET request to grab the JSON data from the weather.json file in the wwwroot folder, the data is rendered within the component.

- `Index.razor`: This is the default component that you will see when you run the project. It shows you how to call other shared components such as `SurveryPrompt` and pass the `Title` parameter to it.

The Shared folder

As the name indicates, this folder holds some shared components across the app. You can override this structure, but we are going to follow the same structure through the rest of this book and will use this folder to contain the components that will be shared across all the pages (app layout, nav menu, authentication components, etc.). By default, the Blazor template comes with three shared components:

- `MainLayout.razor`: This comprises the default layout of the project and is a special kind of component. It defines what the app looks like and allocates the content components (pages) to the right place.

  By default, this component references the `NavMenu` component and contains a `<main>` HTML tag that renders the `@Body` property. This property holds the content of the pages based on the navigated route (more about layout components will be covered in *Chapter 3*, *Developing Advanced Components in Blazor*).

- `NavMenu.razor`: This component contains the menu items of the app (a set of URLs to navigate between the pages mentioned in the `Pages` folder).

- `SuveryPrompt.razor`: This is a simple component to demonstrate how to create a reusable component that accepts parameters. The `Index` page references this component and passes a value for its `Title` parameter.

The _Imports.razor file

In C# classes, you always need to use classes existing in different namespaces. So, basically, we reference a namespace by declaring a `using` statement at the top of the C# file so we can reference the required classes and nested namespaces. The `_Imports.razor` file is there for this purpose, and you can use it to reference common namespaces across the majority of your components. So, there is no need to add a `using` statement at the top of each Razor file to access the required code. By default, the component references some common .NET namespaces that you will use across your components, such as `System.Net.Http` and the namespace of your assembly and the shared folder.

The following is the code for the `_imports.razor` file with the shared referenced namespaces:

```
@using System.Net.Http
...
@using Microsoft.JSInterop
@using BooksStore
@using BooksStore.Shared
```

The App.razor file

The `App` component is the parent and the main component of the Blazor app. It basically defines the infrastructure required for your application to function, such as the `Router` component, which matches the URL entered in the browser and renders the associated component in your app. The `App` component also defines the default layout component (`MainLayout`, mentioned in the `Shared` folder). More about layout components will be covered in *Part 2*, *Chapter 3*, *Developing Advanced Components in Blazor*. Further, the `App` component contains the `NotFound` section, which you can use to render specific content if the requested address was not found.

The Program.cs file

In any C# application, there is an entry point and `Program.cs` represents the entry point of your Blazor app. This entry point sets up the Blazor WebAssembly host and maps the `App` component to the corresponding div within the `index.html` file (by default, the `div` with the ID `app`). We also use the file to register any service that we need to use throughout the application's lifetime in the DI container.

Now you should be able to navigate easily within the solution after understanding what every file does and why it's there; the next step in this chapter is looking at the concept of DI.

Dependency injection in Blazor WebAssembly

Modern software development is all about scaling, separation, and testing, so when you write the code, you should feel confident about its behavior and reliability. To achieve that goal, you need to keep the **SOLID** principles in mind throughout your development journey. The principle represented by the letter *D* is dependency inversion. This basically refers to hiding the implementation of your services behind interfaces. For example, for each service class you have, you can create an interface that contains the methods of that class, and make the class implement the interface. Then, while consuming this service, you reference the interface instead of the class directly. This strategy decouples the components or code pieces that use a service from its implementation by depending on the service's abstract layer, which is the interface.

All this is amazing, but there is one missing part: how to initialize the object of a service that will be used in more than one place. This is where DI comes into play. It's simply a technique that allows you to register your services that are used in other services or components in a centralized container, and that container will be responsible for serving those objects to the components that require them.

How dependency injection works

The following diagram shows the flow of a `ConsoleLoggingService` class:

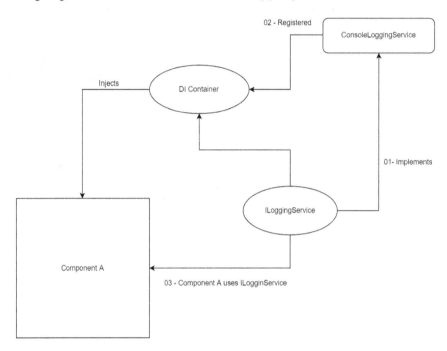

Figure 1.7 – The flow of the dependency injection container and associated services

In the preceding figure, the `ConsoleLoggingService` class implements the `ILoggingService` interface. Then, in the DI container, we register an instance of `ILoggingService` with a new object of its implementation: `ConsoleLoggingService`. Whenever *Component A* requires logging logic, it uses `ILoggingService` (the abstraction layer of `ConsoleLoggingService`).

Instead of initializing a new object instance ourselves, the DI container is serving `ILoggingService` to *Component A*, which adds many benefits as it keeps the code separated. Further, the implementation logic for the full service could be changed at any time without touching *Component A*, which depends on the base interface of that service. For example, if we want to log the data to a server instead of the console to start using the new service, we can just write a new implementation of `ILoggingService` and register that instance in the DI container, without changing anything in *Component A* or any other client code.

> **Note**
>
> To learn more about the SOLID principles and software development, check out the following link: `https://en.wikipedia.org/wiki/SOLID`

Using dependency injection in Blazor WebAssembly

Blazor WebAssembly comes with a DI container out of the box. You can start by registering your services in the `Program.cs` file and then inject those services into the client components or services as well.

Let's implement or write our first service, `ILoggingService`, as mentioned in the preceding section, with the `Log` method, and create an implementation to log the messages to the console window of the browser. Then, we will inject this service into the `FetchData` component and log the count of the weather data that the `FetchData` component gets. To implement the service, go through the following steps:

1. In **Solution Explorer**, right-click on the project, and click on **Add | New Folder**. Name the folder `Services`.

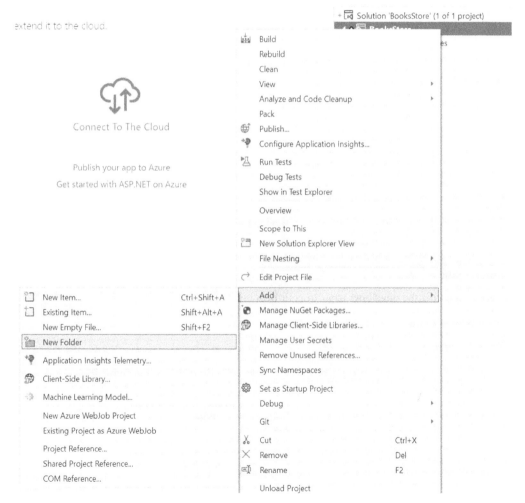

Figure 1.8 – Adding a new folder through Solution Explorer

2. Right-click on the newly created folder and choose **Add | New Item**.

3. From the dialog that shows up, choose **Interface** as the item type, and give it the name `ILoggingService`. Then click **Add**.

4. The interface has been created, and the goal is to create the `Log` method. As you know, in the interfaces, we only define the signature of the method (its return data type and its parameter), so we will define that method as follows:

```
public interface ILoggingService
{
    void Log(string message);
}
```

5. After creating the interface, we should create the implementation that logs that message to the console, so repeat step *2*, but instead of creating an interface, let's create a `class` and give it the name `ConsoleLoggingService`.

6. After creating the class, let's implement the `ILoggingService` interface and write the logic of that method as follows:

```
public class ConsoleLoggingService : ILoggingService
{
    public void Log(string message)
    {
        Console.WriteLine(message);
    }
}
```

The `Log` method calls the `WriteLine` method in the `Console` class, but in Blazor WebAssembly the `Console.WriteLine` method doesn't act the same as it does in other .NET apps by printing a string in a console app. It prints the string in the console window within the developer tools of the browser.

You can access the developer tools in your browser as follows:

- Microsoft Edge: *F12* in Windows or ⌘ + ⌥ + *I* for Mac

- Other browsers: *Ctrl + Shift + I*

The last thing to get this service ready to be injected into other components is registering it in the DI container.

7. Go to the `Program.cs` file and register the service using the `AddScoped` method:

```
...
using BooksStore.Services;
var builder = WebAssemblyHostBuilder.CreateDefault(args);
...
builder.Services.AddScoped<ILoggingService,
```

```
ConsoleLoggingService>();
await builder.Build().RunAsync();
```

Our first service right now is ready to be used and injected within any other components. The following example will show you how to inject this service in the `FetchData` component and achieve the required target of logging the count of weather forecast items to the console window of the browser:

1. Open the `_Imports.razor` file and add a `using` statement to the `Services` namespaces:

    ```
    @using BooksStore.Services
    ```

    We add references to the `Services` namespace in the imports because those services will mostly be used across many components, so we don't have to add that `using` statement in every component.

    Open the `FetchData` component in the `Pages` folder and inject `ILoggingService` by using the `@inject` Razor directive in the component:

    ```
    @page "/fetchdata"
    ...
    @inject ILoggingService LoggingService
    <PageTitle>Weather forecast</PageTitle>
    <h1>Weather forecast</h1>
    ....
    ```

2. Our service is ready to be used and we can call the `Log` function from the object instance in the C# code to log the count of the items as follows:

    ```
        ....
                </tbody>
            </table>
    }
    @code {
        private WeatherForecast[]? forecasts;
        protected override async Task OnInitializedAsync()
        {
            forecasts = await
              Http.GetFromJsonAsync<WeatherForecast[]>
                ("sample-data/weather.json");
            LoggingService.Log($"Number of items retrieved
                            is {forecasts.Count()}");
        }
        ...
    ```

If you run the project and navigate to the **Fetch data** page after the data is retrieved, you can open the console window in your browser and you should see the message **Number of items retrieved is 5**.

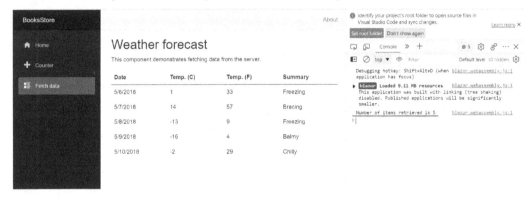

Figure 1.9 – A screenshot of the console window with the printed sentence

> **Tip**
> To learn more about the DI concept in Blazor, you can check the following link: `https://learn.microsoft.com/en-us/aspnet/core/blazor/fundamentals/dependency-injection?view=aspnetcore-7.0`.

Now we have implemented a full cycle, from creating the service abstract layer to its implementation, registering the service in the DI container, and finally injecting and consuming that service from a separate component.

Learning about DI and how it works will help you understand many aspects of the application, such as writing maintainable, decoupled services. Also, you will be able to consume built-in services such as the `HttpClient` to make API calls and the `IConfiguration` service, which we are going to use in the next section to read and consume the application configurations.

Creating, storing, and retrieving the app configurations

Configurations are a set of settings that your application uses throughout its lifetime. App configurations help you avoid hard-coding some values within your app that could change from time to time. Also, the process is very important when dealing with multiple environments. For example, you can store the URL of the API that your app is communicating with. While on the dev machine, the URL is the localhost, in the production environment it refers to the online hosted API (more about environments will be covered in the next section).

Blazor WebAssembly already supports configuration out of the box from JSON files that could be created in the wwwroot folder. The file must be called appsettings.json and to have a configuration file for each environment, you can specify further, for example, appsettings.{ENVIRONMENT}.json or appsettings.Development.json.

Important note

You can add additional sources for configuration but appsettings files are enough as the Blazor WebAssembly app is running fully on the client side. Connecting the app, for example, to the Azure Key Vault service means the connection string will live on the client and be under threat of being exposed easily.

So, for general configurations, appsettings is a great choice. Other kinds of secrets are highly recommended to be stored server-side and we should architect our app in a way that the client doesn't need to fetch or use them.

We will create a configuration file and store some settings such as the URL of the API that we will use. Then, we will print that value with the Index component. Let's get started:

1. Right-click on the wwwroot folder and add a new item of type JSON file and call it appsettings.json.

2. Add a property called "ApiUrl" and give it a value of "http://localhost:44332/" as follows:

    ```
    {
        "ApiUrl": "http://localhost:44332"
    }
    ```

After saving the file, you can access this setting value directly using the IConfiguration service by injecting it into any component or service and referring to the index of your property name.

In the following example, we will inject the IConfiguration service in the Index component and print the value of the ApiUrl property:

1. Open the Index.razor file within the Pages folder and inject the IConfiguration service as shown:

    ```
    @page "/"
    @inject IConfiguration Configuration
    <PageTitle>Index</PageTitle>
    ...
    ```

2. After the service is injected, you are ready to access the required value by using the indexer of the `Configuration` instance. We are going to add a `<p>` tag and display the following: **Api Url: {API_URL_VALUE_IN_SETTINGS}**:

```
...
<h1>Hello, world!</h1>
Welcome to your new app.
<p>Api Url: @Configuration["ApiUrl"]</p>
...
```

After running the application, you should see the following result in your browser window:

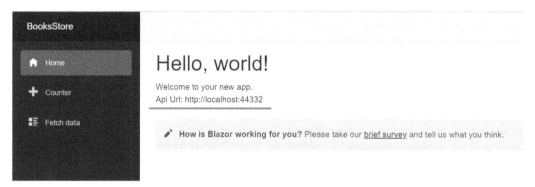

Figure 1.10 – The value of the configuration ApiUrl from appsettings.json

Configurations are important to you in your learning journey, and while developing our real-world application we are going to use them from time to time. After the example we have given, you should have a clear understanding of how to add configurations and how to retrieve their values in the runtime of the app.

In the next section, we will be introducing new environments to our application, and we will add a new configuration file that we will use only while running the app locally on the dev machine using the `dotnet run` command or by debugging the app from within VS 22.

Managing application environments

Software development goes through various stages. In the beginning, we may start with pure development and testing, but after we start rolling out our application to production, many challenges will appear. One of these challenges is having different environments for your app run to on.

For example, imagine your application is communicating with an API and this API is developed by another developer or team. The application is hosted on Azure App Service with two deployment slots (Dev/Production). During development, you are going to use the URL of the dev API, which won't make changes to the production database, and after committing your changes, in the production environment, you need to use the URL of the Production slot.

In this scenario, it will be tough to change the API URL before you push to production and then retrieve it after finishing. Thus, many mistakes may be made, which will make the development process frustrating. This is where managing environments comes into play, where you can write code, render a specific UI, or retrieve special configurations based on the environment that your app is running in without making too many mistakes that delay development and production.

By default, when you run the Blazor app through the debugger, it runs within the development environment, and after publishing it, it runs within the production one.

In this section, we will see how we can set up and retrieve the current environment of the application and retrieve the configuration based on that.

Creating a configuration file based on the environment

To create a configuration file that will be used only in the development phase, you can create another `appsettings.json` file but with the name of the environment before the extension, such as `appsettings.Development.json`. To implement this, take the following steps:

1. Right-click on the wwwroot folder and click on **Add | New Item**. Choose a JSON file and give it the name `appsettings.Development.json`.

2. Add the same `"ApiUrl"` property that exists in `appsettings.json` and the value `"Development"` as follows:

    ```
    {
        "ApiUrl": "Development"
    }
    ```

When you run the application, you will notice that the value of `Development` will show up on the screen instead of `http://localhost:44332`. You should follow the practice of separating the settings based on the environment from the very beginning, so it makes the development process smoother and easier.

Reading the environment within the components

In your development process and where there are many environments for your app (development, preview, and production, for example), you need a specific piece of UI or certain logic to make the app behave differently in different environments. A major implementation to achieve this kind of responsiveness is showing a specific feature in preview but hiding it in production.

To achieve the mentioned behavior, you need to fetch the value of the current environment in your code and use an `if` condition to determine how the UI or logic behaves within that environment.

In the following example, we are going to use `IWebAssemblyHostEnvironment` by injecting it within the `Index` component and remove the `SurveyPrompt` component if the app is running on the development environment:

1. Open the `Index` component within the `Pages` folder and inject the `IWebAssemblyHostEnvironment` service within the component, but of course, you need to reference the namespace that contains the service, which is `Microsoft.AspNetCore.Components.WebAssembly.Hosting`, as shown in the following snippet:

```
@page "/"
@using Microsoft.AspNetCore.Components.WebAssembly.Hosting
. . .
@inject IWebAssemblyHostEnvironment Host
<PageTitle>Index</PageTitle>
. . .
```

2. The next step is to use the `IsDevelopment` method, which returns a bool if the current environment is equal to `Development`:

```
. . .
<p>Api Url: @Configuration["ApiUrl"]</p>

@if (!Host.IsDevelopment())
{
    <SurveyPrompt
        Title="How is Blazor working for you?" />
}
```

After running the application on your machine in debugging mode, the `SuveryPrompt` component will be shown in the production environment. This example demonstrates the responsiveness of environment features, which is very useful.

> **Tip**
> `IWebAssemblyHostEnvironment` contains multiple methods like `IsDevelopment()`. It also contains `IsStaging()` and `IsProduction()`. You can also customize the name by using `IsEnvionment("CUSTOM_ENVIORNEMNT_NAME")`.

One of the topics to be covered is setting the current environment explicitly. We are going to cover this in *Chapter 14, Publishing Blazor WebAssembly Apps*, after publishing the project, but you can go deeper by reading and discovering more about environments at https://learn.microsoft.com/en-us/aspnet/core/blazor/fundamentals/environments?view=aspnetcore-7.0.

Summary

Over the course of this chapter, we have learned the fundamentals of each Blazor WebAssembly application, from setting up the development tools to creating the first application using the *.NET CLI* and *Visual Studio 2022*. We also discovered the anatomy and the structure of the application by defining the purpose of each file and why it's there. We looked at DI, created our first service, injected it within a component, and explored the reasons why it's important in modern software development.

Configurations are an essential part of each application. To avoid hard-coded values within your code, you need to keep the environment that your application is running on in consideration. This extra care can make your development experience easier, especially after publishing an application to production while other features are still under development. Managing the environments and configurations helps with overcoming inconsistencies in development and managing the part under development.

In the next chapter, you will start the real action with Blazor by exploring and developing the basics of app components, which are the main building blocks of a Blazor app.

Further reading

- Application structure: `https://learn.microsoft.com/en-us/aspnet/core/blazor/project-structure?view=aspnetcore-7.0`

- Dependency injection: `https://learn.microsoft.com/en-us/aspnet/core/blazor/fundamentals/dependency-injection?view=aspnetcore-7.0`

- Configuration in Blazor: `https://learn.microsoft.com/en-us/aspnet/core/blazor/fundamentals/configuration?view=aspnetcore-7.0`

- Application environments in Blazor WebAssembly: `https://learn.microsoft.com/en-us/aspnet/core/blazor/fundamentals/environments?view=aspnetcore-7.0`

2

Components in Blazor

Components are the main building blocks of each Blazor application. Each component is a self-contained piece of the **user interface** (**UI**) and its related logic, and comprises a dynamic part of the application UI. A Blazor application is basically a set of components placed together, each of which has its own responsibility within the UI, and the interaction of these components builds up the full app that we are aiming for in this book.

In this chapter, we are going to understand the concept of components from scratch, create our first component, and then use it. We will discover the available mechanisms to pass data between components by introducing **parameters**, **cascading parameters**, and **event callbacks**.

After that, we will look at the component life cycle events and how we can leverage them during the component's lifetime. Finally, we will add some CSS styling for a better look.

In this chapter, we will cover the following topics:

- Understanding the concept of components
- Moving data among components
- Discovering the component life cycle
- Styling the components using CSS

Technical requirements

You can find the source code for this chapter at `https://github.com/PacktPublishing/Mastering-Blazor-WebAssembly/tree/main/Chapter_02`.

Understanding the concept of components

A Blazor application basically consists of a set of components working together to form rich dynamic applications. Blazor components are formally known as Razor components in which **C#** and **HTML** code are combined and the file extension `.razor` is used. Components can be nested within each other, shared, and reused by a different project.

While developing your Blazor application, you should think about it as an organization, where each component is an employee responsible for a certain task. The collaboration of those employees together is what represents the workflow of the organization. The same concept applies to a Blazor app in which every component should have a clear and specific role in the app UI. At the same time, each component should be able to receive corresponding data from other components to get its part of work done.

The following screenshot shows a sample podcast application (*Productive+*, which I developed for tracking how I use my time) built with Blazor:

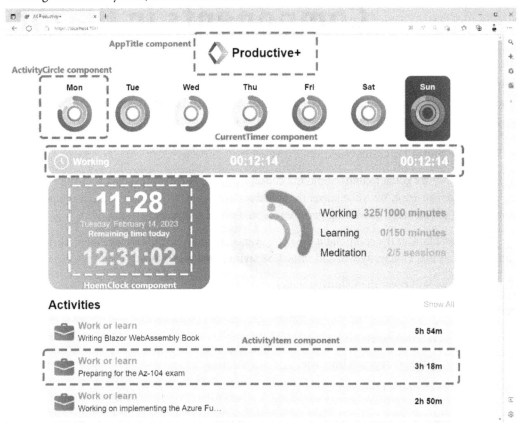

Figure 2.1 – An application I have built, which is split into small components that make up the full UI.
The source code for this sample app is available at https://github.com/aksoftware98/productivity-plus

The preceding figure demonstrates how a specific application UI contains multiple components that together provide us with an advanced, rich UI. The page itself is also a component with nested components contained within it. Finally, all components are hosted together within the `App.razor` component, which is the root of each application.

Now, before we get started creating our first component, let's understand the Razor file and syntax that we use to build the components.

Introduction to Razor

Blazor components are Razor files that give us a super powerful engine to build absolutely dynamic and interactive pieces of UI.

Component files must have a name starting with an uppercase letter, for example, `AlertMessage`.`razor`. If you instead name a component `alertMessage.razor`, you will face a compile-time error, which will prevent the compiler from building the application.

If we look at the pre-built `Counter.razor` component that comes with new Blazor projects (the component results in a default page existing in every new Blazor app and is called *Counter*) in the following code snippet, we can notice the syntax of the Razor component, in addition to the mix of the C# code and HTML:

```
@page "/counter"
<PageTitle>Counter</PageTitle>
<h1>Counter</h1>
<p role="status">Current count: @currentCount</p>
<button class="btn btn-primary" @onclick="IncrementCount">Click me</
button>
@code {
    private int currentCount = 0;
    private void IncrementCount()
    {
        currentCount++;
    }
}
```

As you may notice, the `Counter` component is basically written using the normal HTML tags in addition to a piece of C# code that contains a method called `IncrementCount()`, and a private member called `currentCount`.

In the paragraph tag that comes with the component, the project references the C# `private int` member in the text using the @ char as a prefix. The same applies to the `onclick` event of the button by referencing the method to increment the count of that variable.

Getting started with building components using Razor is a straightforward process as the code doesn't include any complicated sections or initializations; just design your UI using HTML alongside your C# code. It really is as simple as that.

However, before we get started creating our first component, let's introduce the syntax and the available ways to write C# expressions combined with HTML code. In addition to defining some Razor-reserved keywords, we can define some specifications for our components through Razor directives.

Razor syntax

Let's deep dive into Razor syntax and discover the available Razor expressions that we will use to write our components' code:

- **Implicit Razor expressions**: These expressions comprise the first and simplest expressions to employ C# code between HTML tags: constructors from the @ symbol followed by a C# member. After compiling the project, the result of this expression will be rendered and shown in the HTML.

  The following is an example from our Counter.razor component:

  ```
  <p role="status">Current count: @currentCount</p>
  ```

  In the preceding code, we can see the expression with the currentCount variable in the paragraph tag.

  The evaluation of the previous line is that the value of the variable will be rendered in the HTML tag as follows:

  ```
  <p role="status">Current count: 0 </p>
  ```

- **Explicit Razor expressions**: The syntax of these expressions gives you more flexibility to combine complex C# expressions. They use the syntax of the @ symbol followed by two parentheses (..) that contain the C# expressions to be evaluated and rendered.

  The following example shows the evaluation of the currentCount variable and whether it's an even number as a Boolean result (True, False):

  ```
  <p>Is Even: @(currentCount % 2 == 0)</p>
  ```

  The rendering of the previous line, if the value of the currentCount variable was 4, would produce the following HTML:

  ```
  <p>Is Even: True </p>
  ```

- **C# code blocks**: This type allows you to write blocks of C# code for your Blazor component, and the code inside won't be rendered into HTML. Instead, the code will be declared and executed within a component object.

  The syntax of the C# code blocks includes the @ symbol in addition to the code keyword followed by two curly brackets { } that contain the C# code. Back to the example of the Counter.razor, we can see how the code block is being used to create a variable and a function to manipulate the value of that variable as follows:

  ```
  @code {
      private int currentCount = 0;
      private void IncrementCount()
      {
  ```

```
                currentCount++;
        }
    }
```

Razor directives

Behind the scenes, each Razor component is a C# class that gets generated by the Razor engine after parsing it. Razor has what are called directives to provide us with a technique to manipulate the process of this parsing and give some control over some functionalities or the specifications of that generated class.

Razor has two kinds of directives that are divided based on how we can apply them:

- **Directives**: These help manipulate the parsing, specifications, and functionality of the component, and are applied to the component overall

- **Directive attributes**: These directives are applied to some tags and provide some functionality to and manipulation of that specific element

Let's go over some of the most-used directives in Razor components in the following table and define the purpose of each:

Directive	Usage
@page	Makes the component routable by registering it in the router of the application and makes it accessible through a specific URL when requested in the browser through: @page "/counter" We will discuss this further in *Chapter 4, Navigation and Routing*.
@namespace	Enables you to set an explicit namespace for your component through: @namespace BooksStore.Client.Components By default, the namespace of the component is the name of the project followed by the name of the folders it exists inside.
@inherits	Makes the component class inherit from a specific type (implementing inheritance for your component) through: @inherits MyBaseComponent
@inject	Injects a specific service in the component from the DI container through: @inject ILoggingService LoggingService

Directive	Usage

@code	Defines a C# code block:
	
	``
	@code
	{
	    int currentCount = 0;
	    // More C# Code goes here
	}
	``

@using	References a namespace in your component using: ``` @using System.Linq ```
@layout	Defines the layout that will be used for your component: ``` @layout BlankLayout ``` We will discuss this further in *Chapter 3, Developing Advanced Components in Blazor*.
@bind	Implements data binding: ``` <input @bind-Value="currentValue" /> ``` We will discuss this in *Chapter 3, Developing Advanced Components in Blazor*.
@implements	Implements an interface: ``` @implements IDisposable ```

> **Tip**
>
> Razor components support *partial classes*. Razor components are basically C# classes, so you can have the HTML, markup, and C# logic in the same file, or you can have your markup in the .razor file and the C# code in a separate C# class file with the component's name and the same base type, but with the partial keyword, which allows you to divide the same C# class into many files.
>
> This practice is recommended when the C# logic of the component is huge as it helps keep your files small and more organized in addition to leveraging the full power of the C# editor in your IDE.
>
> To learn more about *partial classes* support in Blazor, see https://docs.microsoft.com/en-us/aspnet/core/blazor/components/?view=aspnetcore-7.0#partial-class-support.

Building your first Razor component

After introducing the concept of components and learning the fundamentals of a Blazor component file and its syntax, it's time to put what we have learned into action.

We are going to create our first component, which will be a book card to display a book on our application. We will start from scratch, and then add more functionalities and styling on the go throughout the chapter. So, let's get started:

1. Create a Razor component file in the Shared folder by right-clicking on the *folder* in Visual Studio. Choose **Add | Razor component**, give it the name BookCard.razor, and then click **Add**.

    We have created the component in the Shared folder because it is a shared component and will be used across multiple pages.

2. After you create the file, you are ready to start writing your first markup.

3. We will write a basic HTML design for a card that contains some info about the book, and the button to add the book to the cart, as follows:

    ```
    <div style="border:1px solid black; padding:3px">
        <h6>Book 01</h6>
        <p>Author: Author Name</p>
        <p>Publishing date: 2022-07-02</p>
        <button style="width:100%">Add to Cart</button>
    </div>
    ```

    Now we have a fixed piece of HTML, but this can be reused in many components, and that's what we are going to achieve in the last step.

4. To render a component in another component, all we need to do is reference the name of that component within double tags, < >. To achieve that in our example, let's open the Index. razor component within the Pages folder and reference our book card there.

    The updated file should look like the following:

    ```
    ...
    <PageTitle>Index</PageTitle>

    <h1>Hello, world!</h1>
    Welcome to your new app.

    <BookCard />
    ...
    ```

Now we are ready to see our first reusable Razor component by running the project using the **start** button in Visual Studio. The **Index component** will automatically open as it's the default component, and we should see our book card there, as shown in the following screenshot:

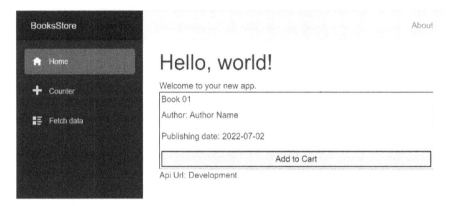

Figure 2.2 – The result of the rendered BookCard component on the Index page

Congratulations! You have successfully created and used your first Razor component after learning the basics of components in general and Razor specifically. We can now take a step ahead by making our component a bit dynamic via the different methods to send data from the parent components to the children, and in the opposite direction, by using `Parameters`, `CascadingParameters`, and `EventCallbacks`.

Moving data among components

Methods in C# work by accepting parameters to process and return other types of data post processing. Blazor components work according to the exact same concept. Because each component represents a piece of UI alongside its logic, there is always the need to give the component some data for it to function according to the set logic. So functionally, a Blazor component either renders the data that is input in a certain way, or responds based on the set logic.

A Blazor component, in some cases, requires some data as an input either to render the data in the UI or to control its internal logic based on the values provided. Also, in many scenarios, components have the ability to send data back to the parent. The effective collaboration of the app components by communicating with each other provides us with the clean, well-functioning, and effective app that we are trying to build.

Luckily, Blazor provides us with powerful mechanisms to allow the transfer of data between components and in different directions (from parent components to children, and vice versa).

Now it's time to discover these mechanisms and apply them to the component we created in the preceding section.

Parameters

A Blazor component can accept parameters just like any C# method, but the definition of parameters in Blazor is way different. **Parameters** in Blazor components are C# properties decorated with the [Parameter] attribute, and we then set their values by passing them as attributes while calling the component.

So, let's get started adding some parameters to our BookCard component to parametrize the data of it by accepting this data as parameters instead of hardcoded values in the HTML.

To achieve this, we need to add the @code section to our component and define the following parameters (Title, PublishingDate, and Author), but the properties of the object (book) could change from time to time during the application development cycle. So instead of passing each property as a parameter, we can create a Book class and then pass the book object to the component.

This practice will help the maintainability of your Blazor components because when you add new properties to your model, you don't need to add the corresponding parameter to the component or even to multiple components using the same model. There is also no need to add those new properties as attributes when calling the component. To achieve all of this, we need to do the following:

1. Right-click on the project, then choose **Add | New Folder** and give it the name Models; the folder will contain the models that we will use in our project.

2. Add a new C# class called Book within the **Models** folder by right-clicking on the folder, choosing **Add | New Item**, adding a C# class, and giving it the name Book.cs.

3. Next, we need to add the properties of the book to the created class. The properties we need are Title, PublishingDate, and AuthorName, so the class file should look like this:

```
namespace BooksStore.Models;
public class Book
{
    public string? Title { get; set; }
    public string? AuthorName { get; set; }
    public DateTime PublishingDate { get; set; }
}
```

4. Before we move to the BookCard component, we just need to add the namespace BooksStore. Models to the _Imports.razor file so we don't need to add the using statement every time we want to reference a model:

    ```
    . . .
    @using BooksStore.Shared
    @using BooksStore.Services
    @using BooksStore.Models
    ```

5. Open the BookCard.razor file, and let's modify it to accept a Book object as a parameter by adding the @code { } directive and parameter of the type Book. The following code snippet shows the result of this step:

    ```
    . . .
        <button style="width:100%">Add to Cart</button>
    </div>

    @code
    {
        [Parameter]
        public Book? Book { get; set; }
    }
    ```

6. After adding the parameter, we can reference the value of the properties of the Book object using the implicit Razor expressions in HTML via @ and the name of the property within the parameter object (we will also use the *nullable operator* (?) just in case the parameter value was null) as the next code snippet shows:

    ```
    <div style="border:1px solid black; padding:3px">
        <h6>@Book?.Title</h6>
        <p>Author: @Book?.AuthorName</p>
        <p>Publishing date: @Book?.PublishingDate</p>
        <button style="width:100%">Add to Cart</button>
    </div>
    . . .
    ```

7. The last step is to provide the value of the Book parameter to the component when we call it. Go back to the Index.razor file and add a new @code {} directive that will include the declaration of a new Book object:

    ```
    . . .
    @if (!Host.IsDevelopment())
    {
        <SurveyPrompt Title="How is Blazor working for
          you?" />
    }
    ```

```
@code
{
    private Book _firstBook = new Book
        {
            AuthorName = "John Smith",
            PublishingDate = new DateTime(2022, 08,
                                          01),
            Title = "Mastering Blazor WebAssembly"
        };
}
...
```

8. Modify the line where it calls the BookCard component by passing the value of the Book parameter using an attribute:

```
...
<BookCard Book="_firstBook"  />
...
```

So, if we run the project right now, we should see the exact same component but with different data rendered from the previous one. This data is passed using parameters when calling the component, as shown in the following screenshot:

Figure 2.3 – Rendered component after passing the parameters

The same component can now be used just as we have seen all over the application and whenever your app needs to show this piece of UI. The Book parameter is used to pass data to be rendered within the component, but now we will create another parameter that will control the rendering process and the functionality of the component.

The second parameter we will add will be responsible for showing or hiding the **Add to Cart** button as, in some places in the app, we just need to see the info of the book without the need to be able to add it to the shopping cart.

To achieve this behavior, we will create another parameter of the `Boolean` type called `WithButton` and directly initialize the `True` value in the `BookCard` component, as the following code shows:

```
. . .
[Parameter]
public bool WithButton { get; set; } = true;
. . .
```

Now, after creating the parameter, we can wrap the button with an `if` condition to show it only if the value of this button is true, as in the following code snippet:

```
. . .
<p>Publishing date: @Book?.PublishingDate</p>

    @if (WithButton)
    {
        <button style="width:100%">Add to Cart</button>
    }
</div>
. . .
```

Now we are going to call this component one more time in the `Index` component in the `Index.razor` file, but providing the `false` value to `WithButton`:

```
. . .
<BookCard Book="_firstBook" />

<BookCard Book="_firstBook" WithButton="false" />

<p>Api Url: @Configuration["ApiUrl"]</p>
. . .
```

You may notice, after running the application one more time, that the book card is present twice, but the second one doesn't contain an **Add to Cart** button.

Figure 2.4 – Rendering two book cards, the second one without a button

Parameters – special cases

Blazor also has some other complementary attributes that provide some additional functionality and define some specifications for the parameters of your components. Here are some special cases and attributes that you will mostly use while developing your application:

- **[EditorRequired] attribute**: If you decorate a Blazor parameter with this attribute, the parameter will be required, and the compiler will show some warnings if the value of the parameter is not provided.

  We can make our Book parameter required in the BookCard component by adding it as follows:

  ```
  ...
  @code
  {
      [Parameter]
      [EditorRequired]
      public Book? Book { get; set; }
  }
  ```

  Or it could be added in the same line:

  ```
  [Parameter, EditorRequired]
  public Book? Book { get; set; }
  ```

- **Capture arbitrary attributes**: Sometimes you need to pass some attributes to your components that are not declared as parameters. This is most likely the case when you want to capture the HTML element attributes such as class, style, and so on.

You can capture those undefined attributes and assign them to an element in your component using the @attributes attribute for your HTML tag.

The value of this attribute is Dictionary<string, object>. The key of each dictionary element represents the name, and the value of type object represents the value of that property.

Back in our BookCard component, we want to improve our component by adding the ability to capture attributes such as style and class, and apply them to the parent div of the card.

To achieve this, in BookCard.razor, we need to create a parameter called UserAttributes of the Dictionary<string, object>type, and set the CaptureUnmatchedValues = true property for the Parameter attribute, as the following code snippet shows:

```
...
    [Parameter(CaptureUnmatchedValues = true)]
    public Dictionary<string, object>? UserAttributes
      { get; set; }
...
```

Now we can assign the value of this parameter to the @attributes directive in the div tag of the card as follows:

```
...
<div style ="border:1px solid black; padding:5px" @
attributes="UserAttribues">
    <h6>@Book?.Title</h6>
    ...
```

Now while calling the component, we can pass attributes such as class and onclick to the parent div without defining them as parameters explicitly.

As you have noticed in the preceding example when we added another BookCard, there is no margin between the two cards in the Index component. We will add some margin by adding the class attribute and giving it the bootstrap class mt-3, which will add some margin from the top as follows:

```
<BookCard Book="_firstBook" />
<BookCard Book="_firstBook" WithButton="false" class="mt-3"/>
```

The class attribute now will be applied to the div wrapper of the card and will add a margin at the top:

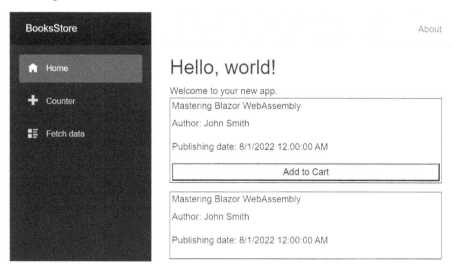

Figure 2.6 – The second book card after adding a margin at the top

EventCallback

EventCallback is another smart technique provided by Blazor that allows the child component to send data to its parent by assigning a function to the child component that it will call when a certain event occurs.

The actual value of EventCallback is a function that either takes no or one parameter if we use the generic version of it (EventCallback<T>).

The functions to be used as an EventCallback value should either return void or Task. The technique doesn't support returning other types.

> **Tip**
>
> EventCallback is wrapped around the concept of delegates in C#, with some added functionality to handle the updated state of the component.
>
> A delegate is basically a C# type and its value represents a function with a predefined signature (return type and parameters).
>
> To learn more about delegates, see https://docs.microsoft.com/en-us/dotnet/csharp/programming-guide/delegates/.

Now we will extend the functionality of the `BookCard` by adding an `EventCallback` that takes `Book` as a parameter, and this event will be fired when the user clicks on the **Add to Cart** button.

So, in `BookCard.razor`, to add an event callback with the `Book` type for its parameter, we need to create a Blazor parameter of the `EventCallback<Book>` type, and give it the name `OnAddToCartClicked`:

```
. . .
[Parameter]
public EventCallback<Book?> OnAddToCartClicked { get; set; }
```

The preceding code snippet means the `BookCard` component can accept a function that returns `void` or `Task` and takes a parameter of the `Book` type.

Now to fire this event, we will create a simple function called `AddToCart` that will fire the `EventCallback` and pass the book of the card as a parameter:

```
. . .
[Parameter]
public EventCallback<Book?> OnAddToCartClicked { get; set; }

private void AddToCart()
{
    OnAddToCartClicked.InvokeAsync(Book);
}
. . .
```

To fire the `AddToCart()` function, we need to use the `@onclick` event for the button that will be triggered when we click the **Add to Cart** button as follows:

```
. . .
@if (WithButton)
{
    <button style="width:100%" @onclick="AddToCart">
      Add to Cart</button>
}
. . .
```

After adding this `EventCallback`, the `BookCard` now allows its parents to assign a function that takes `Book` as a parameter. That means when that function in the parent is triggered in `BookCard`, it will pass the `Book` object from the child to the parent via the parameter of the function, where it can do some processing outside the scope of the `BookCard` component that is responsible for rendering the book only.

The next phase requires us to modify `Index.razor` so it will show a list of books instead of one, and when the user clicks **Add to Cart**, the given title will be added to the user's cart in the UI using the following steps:

1. To deal with the books in our project (retrieve, add, edit, etc.), we will create a service with its interface, register it in the DI container, and then inject it whenever we need to.

    To achieve that, create a new interface called `IBooksService` in the `Services` folder and add a method inside it called `GetAllBooksAsync`:

    ```
    using BooksStore.Models;
    namespace BooksStore.Services;
    public interface IBooksService
    {
        Task<List<Book>> GetAllBooksAsync();
    }
    ```

2. Next, also in the `Services` folder, create a new class with the name `LocalBooksService. cs`. The class will have an implementation for the interface we created in *step 1*. The `LocalBooksService` class will provide methods to deal with book objects stored in the local memory as follows:

    ```
    using BooksStore.Models;
    namespace BooksStore.Services;
    public class LocalBooksService : IBooksService
    {
        static List<Book> _allBooks = new List<Book>
        {
            new Book
            {
                AuthorName = "John Smith",
                PublishingDate = new DateTime(2021, 01,
                                             12),
                Title = "Blazor WebAssembly Guide"
            },
            new Book
            {
                AuthorName = "John Smith",
                PublishingDate = new DateTime(2022, 03,
                                             13),
                Title = "Mastering Blazor WebAssembly",
            },
            new Book
            {
                AuthorName = "John Smith",
                PublishingDate = new DateTime(2022, 08,
    ```

```
                                                    01),
                    Title = "Learning Blazor from A to Z"
                }
        };

        public Task<List<Book>> GetAllBooksAsync()
        {
            return Task.FromResult(_allBooks);
        }
    }
```

3. The `GetAllBooksAsync` returns `Task<List<Book>>` because later, in *Chapter 8, Consuming Web APIs from Blazor WebAssembly*, we will create another implementation for `IBooksService` that will require `awaitable` calls for an online service to fetch the books instead of retrieving them from a static in-memory list.

4. To be able to inject `IBooksService`, we need to register it in the DI container with its implementation. So, in `Program.cs`, you can add the following line:

   ```
   builder.Services.AddScoped<IBooksService, LocalBooksService>();
   ```

5. Open the `Index.razor` file and inject `IBooksService` using the `@inject` directive at the top:

   ```
   ...
   @inject IBooksService BooksService
   ```

6. In the `@code` section, create a variable of the `List<Book>` type, override the `OnInitializedAsync` method, and then call `GetAllBooksAsync` from `BooksService` as follows:

   ```
   ...
       private List<Book> _books = new List<Book>();
       protected override async Task OnInitializedAsync()
       {
           _books =
             await BooksService.GetAllBooksAsync();
       }
   ...
   ```

7. To show this list of books, we need to iterate over it using a `foreach` loop and render the `BookCard` component in each iteration. In addition to that, we will create a `div` wrapper with `style` as `display:flex`, so the books will be rendered horizontally:

   ```
   ...
   <BookCard Book="_firstBook" WithButton="false" class="mt-3" />

   <h3>Available Books:</h3>
   <div style="display:flex">
   ```

```
    @foreach (var book in _books)
    {
        <BookCard Book="book" />
    }
</div>
...
```

The result of the previous work should be something like this:

Available Books:

Blazor WebAssembly Guide	Mastering Blazor WebAssembly	Learning Blazor from A to Z
Author: John Smith	Author: John Smith	Author: John Smith
Publishing date: 1/12/2021 12:00:00 AM	Publishing date: 3/13/2022 12:00:00 AM	Publishing date: 8/1/2022 12:00:00 AM
Add to Cart	Add to Cart	Add to Cart

Figure 2.6 – List of books

8. After having the list, we can leverage the `OnAddToCartClicked` event callback. So, create a list of titles in the `Index`, and every time the user clicks **Add to Cart**, the title of that book gets added to that list and rendered in the UI. In the code, create a `List<Book>` object and call it `_booksCart`:

    ```
    private List<Book> _booksCart = new List<Book>();
    ```

9. The following is the books list in the UI. We will iterate over `_booksCart` and print the title of each book in the cart:

    ```
    ...
    <h3 class="mt-3">My Cart</h3>
    <ul>
        @foreach (var item in _booksCart)
        {
            <li>@item.Title</li>
        }
    </ul>
    ...
    ```

    Initially the list will be empty, but in the next step, we will create the functionality to add a book to the cart every time the user clicks **Add to Cart**.

10. Now create a new function called `AddToCart` that takes a `Book` object as a parameter. This function will be passed to the event callback of the `BookCard` component. The `BookCard` will trigger this function and the book of the card will be passed to this function as a parameter so we can process it. Our processing, for now, is adding the book to the cart list:

    ```
    private void AddToCart(Book selectedBook)
    {
    ```

```
        _booksCart.Add(selectedBook);
    }
```

11. The last step is just assigning the `AddToCart` function to the `OnAddToCartClicked` event callback of the `BookCard` component as follows:

```
<div style="display:flex" >
    @foreach (var book in _books)
    {
        <BookCard Book="book"
          OnAddToCartClicked="AddToCart" />
    }
</div>
```

Please note that when you assign the function, don't use the parentheses as in `AddToCart()`, because here we are not calling the function. Instead, we are assigning it just like any other variable.

Now, after running the project and clicking on the **Add to Cart** button below a book in the **Available Books** list, you will notice that the title of the book will be added to the **My Cart** list:

Available Books:

Blazor WebAssembly Guide	Mastering Blazor WebAssembly	Learning Blazor from A to Z
Author: John Smith	Author: John Smith	Author: John Smith
Publishing date: 1/12/2021 12:00:00 AM	Publishing date: 3/13/2022 12:00:00 AM	Publishing date: 8/1/2022 12:00:00 AM
Add to Cart	Add to Cart	Add to Cart

My Cart

- Blazor WebAssembly Guide
- Mastering Blazor WebAssembly

Figure 2.7 – Cart list after clicking on Add to Cart

The power of `EventCallback`, which allows the child component to call a function from the parent component and pass data for it, gives us a very flexible way to communicate while keeping the code separated and clean.

Cascading values and parameters

Cascading values and parameters comprise another way to make data flow from the parent components to their children. However, the difference between cascading parameters and normal component parameters is that cascading parameters pass the data to any number of descendent components in the hierarchy. Unlike the component parameters, cascading values and parameters don't need an assignment through the attribute; their values are populated automatically.

And because the cascading values can go down in the hierarchy of the components, any child component can access a cascading value from a parent component despite the number of levels between them.

The following code snippet is a basic setup for the cascading value:

```
<CascadingValue Value="_cardStyle">
    <BookCard Book="_firstBook" />
</CascadingValue>
```

To start using the cascading values and parameters, we first have the CascadingValue component that allows us to set a cascading value in a parent component, and the [CascadingParameter] attribute is to make use of the cascading value provided by an ancestor component.

In the following example, we are going to use the MainLayout file in the Shared folder to add a cascading value component that wraps the @Body property, which holds all the child components within the layout. (More about layouts will be covered in *Chapter 3, Developing Advanced Components in Blazor*.) The cascading value will be a basic string variable that represents the major style of the background for the interested child components. And that will unify the background style for all the cards that we will have:

1. Open the MainLayout.razor file in the Shared folder, create a string variable called _backgroundStyle, and initialize it with a basic style:

    ```
    . . .
    @code
    {
        private string _backroundStyle =
          "background-color:#f2f2f2";
    }
    ```

2. Within the article HTML tag, wrap the @Body property render with the CascadingValue component and assign the Value property to @_backgroundStyle:

    ```
    . . .
            <article class="content px-4">
                <CascadingValue Value="@_backroundStyle">
                    @Body
                </CascadingValue>
            </article>
    . . .
    ```

    Now all child components have access to a cascading parameter of type string that holds the basic style for the background and could be used by any component within the hierarchy.

3. Back in the `BookCard` component, we need to make use of the value provided by the parent component, which is `MainLayout`. To access that value, you need to add a C# property of type string and decorate it with the `[CascadingParameter]` attribute as follows:

```
. . .
@code
{
    [CascadingParameter]
    public string? BackgroundStyle { get; set; }
. . .
```

4. Now, the value from the main layout will automatically be populated in the `BackgroundStyle` parameter, so we can append it to the `div`'s style attribute as follows:

```
<div style="border:1px solid black; padding:3px;@
BackgroundStyle" @attributes="UserAttributes">
    <h6>@Book?.Title</h6>
. . .
```

Now after running the project, note that all the `BookCard` components on the `Index` page have the same background color, and changing the value in `MainLayout` will update all components using this value.

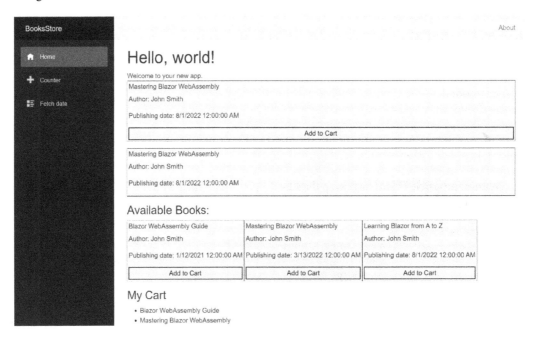

Figure 2.8 – Unified background style for all the BookCard components in the UI

IsFixed parameter for cascading values

The CascadingValue component has an optional parameter called IsFixed of bool type; its value is false by default. IsFixed instructs the recipient components of the cascading value to either subscribe to the value change or not.

It's recommended, when possible, to set its value to true, as that will lead to no subscriptions receiving updates about cascading value changes, resulting in better performance, especially when there are a large number of recipients for the value.

You can set the IsFixed parameter as follows:

```
<CascadingValue IsFixed="true" Value="this">
    <ChildComponent />
</CascadingValue>
```

Multiple cascading values and parameters

As you have seen in the preceding example, we could use the cascading value because it was the only type string available.

To be able to add multiple cascading values, you need to either have multiple values with multiple types (string value, int value, Book value, etc.), or the same value but defined with the name of the value explicitly:

- **Multiple values with different types**: The following examples in the MainLayout component show how we can declare more than one cascading value of different types (string and bool in the following example):

```
    ...
    <article class="content px-4">
            <CascadingValue Value="@_backroundStyle">
                <CascadingValue Value="@_isBusy">
                    @Body
                </CascadingValue>
            </CascadingValue>
        </article>
    </main>
</div>

@code
{
    ...
    private bool _isBusy = false;
}
```

Now, we have two cascading values available in any child component to access by defining their cascading parameters. One is of type bool and the other is of type string as follows:

```
[CascadingParameter]
public string? BackgroundStyle { get; set; }

[CascadingParameter]
public bool IsBusy { get; set; }
```

- **Multiple values with explicit names**: If you have multiple cascading values with the same types, you can define the name of the value explicitly using the Name parameter, as shown in the following example:

```
<article class="content px-4">
    <CascadingValue Value="@_backroundStyle"
      Name="BackgroundStyle">
        <CascadingValue Value="@_buttonStyle"
          Name="ButtonStyle">
            @Body
        </CascadingValue>
    </CascadingValue>
...
@code
{
    private string _backroundStyle =
      "background-color:#f2f2f2";
    private string _buttonStyle =
      "background-color:orange";
```

Now, when referencing one of those cascading values in the child components, we need to define the name using the Name property of the [CascadingParameter] attribute as shown in the BookCard component:

```
[CascadingParameter(Name = "BackgroundStyle")]
public string? BackgroundStyle { get; set; }

  [CascadingParameter(Name = "ButtonStyle")]
public string? ButtonStyle { get; set; }
```

With that, we have discovered three different ways to enable communication between our application components: parameters, cascading parameters, and event callbacks.

Two-way data binding in Blazor

Now we know that we can use parameters to pass data from a parent to its children, and event callbacks to pass data from the child to the parent. Next, we will learn about combining them together.

Blazor provides a neat and compact way to enable two-way data passing between components. Let's take a `Table` component that renders a collection of data as an example. The `Table` component allows the user to select an item when clicking on any row. Basically, in the `Table` component, we need a parameter called `SelectedItem` and an `EventCallback` called `OnSelectedItem` that will enable us to pass an already selected item to the table. The `EventCallback` will also let us receive the new selected item when the user picks it.

To get the update from the child component, we need to create a method and assign it to the `EventCallback` parameter of the component. This method will accept the passed value from the parameter, and then assign it to a variable in the very basic shape if there is no processing required. To send the value of that variable to the component, we need to pass it to the `Parameter`. With the `@bind` directive, we can combine both operations and skip the method that will update the variable in a single attribute.

To be able to use the `@bind` with a component, it has to have a parameter of a certain type and `EventCallback` of the same parameter type. The `EventCallback` must have the same name as the parameter concatenated with the word *Changed*, as the following example shows:

```
[Parameter]
public string? Message { get; set; }
[Parameter]
public EventCallback<string?> MessageChanged { get; set; }
```

That component accepts a parameter of type string, and it also updates the value of that string internally and it exposes an `EventCallback` that could be triggered when that value gets changed. So, when we use this component in another parent component that has a variable of type string called `_message`, this variable will be passed to the child component, and we also need to update it when the child updates its value internally, so let's use `@bind` as follows:

```
<ChildComponent @bind-Message="_message" />
```

In *Chapter 3, Developing Advanced Components in Blazor*, we will use the two-way data binding while developing a component; also in *Chapter 5, Capturing User Input with Forms and Validation*, while using the input components, we will use it heavily and learn more about its advanced techniques.

In the next section, we will discover the life cycle of the component within the Blazor app, and each stage the component goes through over its lifetime.

Discovering the component life cycle

Blazor components go through a set of methods from initialization to rendering and finally being disposed of. These methods have synchronous and asynchronous versions that we can leverage to perform certain tasks.

The following list shows the methods that the components go through; not all of them are being called. Being called depends on whether the component is being rendered for the first time:

- `SetParameterAsync`: This sets the value of the parameters from the component parent and the route parameters. (More about route parameters will be covered in *Chapter 4, Navigation and Routing*.)

    The default implementation of this method sets the values of the parameters and the cascading parameters available. `SetParameterAsync` is called only for the first render of the component. Overriding its implementation allows you to control the process of setting the values of the parameters and write some logic based on that, as shown in the `BookCard.razor` file:

    ```
    // Override the SetParametersAsync to control the
    // process of setting the parameters
    // using the ParameterView object that holds the
    // values of the component parameters
    public override async Task SetParametersAsync(ParameterView
    parameters)
    {
        if (parameters.TryGetValue<string>(nameof(BackgroundStyle),
    out var value))
        {
            if (string.IsNullOrWhiteSpace(value))
            {
                BackgroundStyle = "background-color:white";
            }
        }
        await base.SetParametersAsync(parameters);
    }
    ```

- `OnInitialized`/`OnInitializedAsync`: After the component receives its parameter values, `OnInitialized` for synchronous operations and `OnInitializedAsync` for asynchronous operations are invoked, only for the first-time render.

Overriding the implementation of `OnInitialized` or `OnInitializedAsync` allows us to take some actions and execute logic when the components get initialized, for example, to start fetching data from the API (more on this in *Chapter 8, Consuming Web APIs from Blazor WebAssembly*):

```
protected async override Task OnInitializedAsync()
{
    // Fetch data from the API
}
```

- `OnParameterSet` / `OnParameterSetAsync`: These two methods get called after `OnInitialized` and `OnInitializedAsync` in the first render, as well as after one or more parameter values get changed by the parent component.

  To validate or do some logic when any value of a parameter gets changed, you need to override either `OnParameterSet` or `OnParameterSetAsync` for asynchronous logic:

```
// Validate the Book parameter and throws an exception
// if it's null
protected override void OnParametersSet()
{
    if (Book == null)
        throw new ArgumentNullException(nameof(Book));
}
```

- `OnAfterRender` / `OnAfterRenderAsync`: These methods are called after the component rendering has finished, and any changes here won't be rendered unless you explicitly use the `StateHasChanged()` method. In this stage, all the elements and the component references are fully rendered.

  `OnAfterRender` and `OnAfterRenderAsync` has a Boolean parameter that determines whether this is the first render of the component or not. As these methods get called after the rendering process for the component, the process may comprise the first render or a subsequent one. Overriding those methods is perfect for making JS calls against the DOM (more on this in *Chapter 6, Consuming JavaScript in Blazor*), as the DOM elements are rendered and available:

```
protected async override Task OnAfterRenderAsync(bool
firstRender)
{
    // Make a JavaScript call to manipulate the DOM
    // elements
}
```

The following diagram shows the flow of the component life cycle:

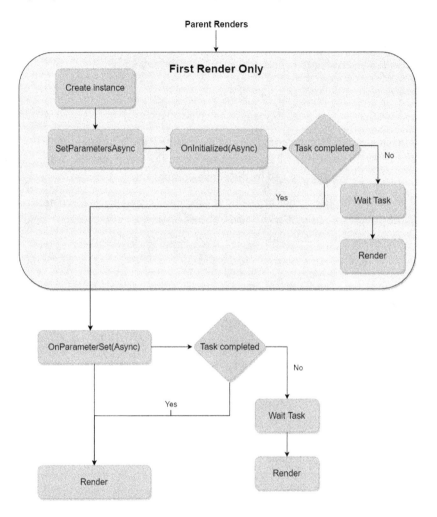

Figure 2.9 – Component lifetime stages

Tip

If the asynchronous task executed in the life cycle events is not completed, the component will continue to be rendered. When the task gets completed, the component will be rendered to reflect the updates in the UI if there are any.

Dispose event

One extra set of life cycle events comprises the `Dispose` and `DisposeAsync` methods by implementing either the `IDisposable` interface for synchronous disposal or `IAsyncDisposable` for asynchronous disposal. Disposal will occur when the component is removed from the UI and helps to clear up unmanaged resources such as files and events.

The following example shows how to implement the `Dispose` method:

```
@implements IDisposable
     ...
@code {
public void Dispose()
{
    // Release unmanaged resources like unsubscribe from
    // events
    // Timer.Dispose();
}
...
}
```

> **Tip**
> You need to implement either `IDisposable` or `IAsyncDisposable` as the framework will call the asynchronous version if both are available.

In this section, we have learned about the life cycle events that each Blazor component has. These events are used heavily in each application and throughout the projects in this book.

Now, after creating a full Blazor component and populating it with some data, it's time to make it pretty by adding some CSS for it.

Styling the components using CSS

CSS is a web utility that allows us to change the look and feel of our elements. In this section, we will deep dive into the different ways supported by Blazor with which we can add CSS styles to our components.

In Blazor apps, we have four different ways to style our components:

- Global styles
- Isolated styles (scoped)
- Inline styles
- Embedded styles

Isolated styles

Blazor provides us with a powerful way to style our components by keeping the styles separated from each other. This means small, clean styles, scoped for each component.

To reference the isolated styles, you need to add a link tag with a reference using the syntax `{AssemblyName}.styles.css`.

Let's get started using isolated styles to style our `BookCard` component:

1. Create a CSS file in the same folder of the `BookCard` component following the convention `{ComponentName}.razor.css` – in our case, `BookCard.razor.css`. This will make Visual Studio create a nested file behind the component file as follows:

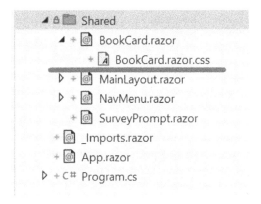

Figure 2.10 – Nested CSS file by Visual Studio

All the styles within `BookCard.razor.css` will be scoped for the `BookCard` component only.

2. Let's add some CSS styles to make the book's card looks prettier. We are going to have two classes: one for the `button` and another for `div` with margin, padding, a little shadow, and some smooth borders for a modern feel:

```css
.card {
    margin: 7px;
    padding: 7px;
    box-shadow: 0px 3px 8px -3px rgb(0 0 0 / 75%);
    border-radius: 3px;
}

.main-button {
    background-color: #ec6611;
    border-radius: 3px;
    border: none;
```

```
        color: white;
        width: 100%;
        padding: 4px;
        margin: 4px;
        font-size: 13px;
        cursor: pointer;
    }
```

3. Back in the `BookCard.razor` file, we can now apply those classes to the `div` and `button` elements and we can remove the inline styles we set earlier:

```
<div style="@BackgroundStyle"     @attributes="UserAttributes"
    class="card">
    <h6>@Book?.Title</h6>

...

    @if (WithButton)
    {
        <button class="main-button"
          @onclick="AddToCart">Add to Cart</button>
    }
...
```

Now when we run the project, we will see that our card has a much better visual appeal than before. The result should be like the following screenshot:

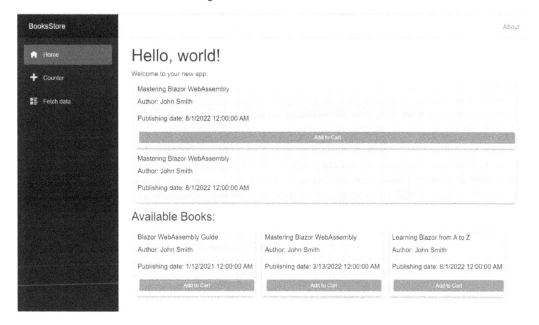

Figure 2.11 – The look of the BookCard after adding CSS

The powerful part of this CSS is that the class names (card and main-button) are only scoped for the BookCard component and cannot affect other components. The concept of isolation here gives us clean and short CSS files, in addition to having the styles of each component alongside the respective component file.

So, now the question is: how does isolated CSS work?

After running the project, open the developer tools and look at the generated HTML and CSS. Note that Blazor has appended an HTML attribute in the format b-{10-characters-text}. The following screenshot shows the div element of the card:

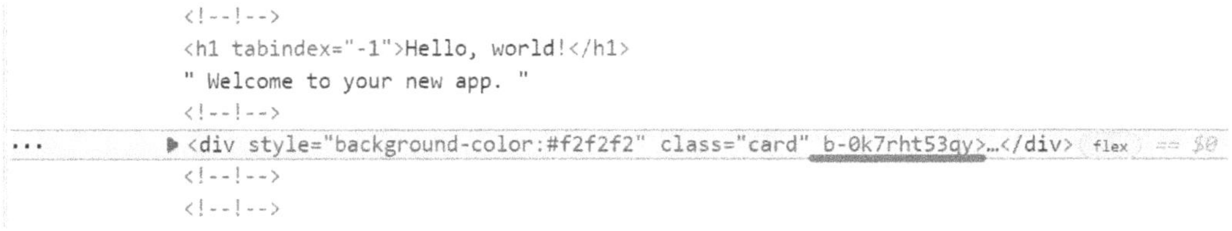

```
<!--!-->
<h1 tabindex="-1">Hello, world!</h1>
" Welcome to your new app. "
<!--!-->
▶ <div style="background-color:#f2f2f2" class="card" b-0k7rht53qy>...</div>  flex   == $0
<!--!-->
<!--!-->
```

Figure 2.12 – The HTML attribute appended to the div element by Blazor

And looking at the following CSS window, we can also see that the card class has been renamed, with the same identifier appended as follows:

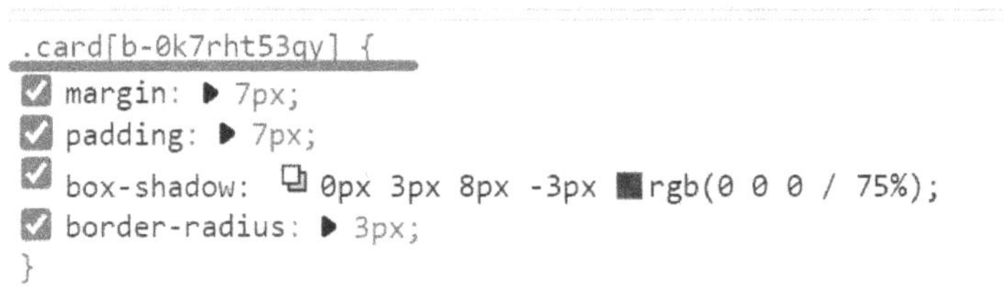

```
.card[b-0k7rht53qy] {
  margin: ▶ 7px;
  padding: ▶ 7px;
  box-shadow: ⊡ 0px 3px 8px -3px ■rgb(0 0 0 / 75%);
  border-radius: ▶ 3px;
}
```

Figure 2.13 – The updated CSS class name with the identifier

In essence, the CSS isolation happens at build time when Blazor rewrites all the CSS selectors to target the HTML elements with the unique identifier. Then, it combines them in a static CSS file with a name using the syntax {AssemblyName}.styles.css. If you view your project in the CSS window of the browser **Dev Tools**, you will see the name of the CSS files that contain classes, such as BooksStore.styles.css:

```
Filter                          :hov  .cls  +  ♎  ⬅

.card[b-0k7rht53qy] {               BooksStore.styles.css:2
  margin: ▶ 7px;
  padding: ▶ 7px;
```

Figure 2.14 – Name of the CSS file that contains the isolated CSS selectors

And that's why if we open the `Index.html` file, we see that Blazor references the `BooksStore.styles.css` file, as this file gets generated at the build time and is available in the root folder:

```
<html lang="en">
<head>
...
    <link href="BooksStore.styles.css" rel="stylesheet" />
</head>
...
```

> **Note**
>
> Isolated CSS will be applied only to the scoped component. You can use the `::deep` pseudo-element, for example, `.card::deep`, so that styles will be applied to descendent elements in the child components.

Global styles

Global styles use a `.css` file for the application referenced in the `Index.html` file, like the default `app.css` file found in the `wwwroot/css` folder. By default, that file has the styles for the default Blazor template. You can use this file to set general styles that will be used across your application, such as text styles.

Inline styles

Inline styles are created when you set the CSS styles directly in the element within the component using the style attribute. This method is efficient as no cache issues will be faced.

However, this will lead to a long element written in the components, and the styles applied to the element cannot be reused because they are not wrapped with a class.

The benefit of using the inline styles is to have more control over the CSS properties based on a certain logic in your component, for example, changing the color of the text if a book is being sold at a discount.

> **Tip**
>
> When working with inline styles, especially those managed by specific behavior, it's recommended to use the third-party `BlazorComponentUtilities` package, which contains the `StyleBuilder` class, and helps us write inline styles using C# in a very clean and manageable way. You can download the package from `https://www.nuget.org/packages/BlazorComponentUtilities/`.

Embedded styles

Embedded styles can be created by using the `<style>` tag within your component, and will override other styles. However, this approach is not recommended because it will lead to repeating the styles every time the component gets rendered or shown in the UI.

The following example shows how you can use embedded styles in your component:

```
...
<style>
    .main-button:hover {
        background-color: #c7540c
    }
</style>
...
```

The embedded-styles approach is useful if you are using a Blazor UI framework and you want to override some of the default styles it has. To do this, open the **Dev Tools** in the browser and discover the class names. Then, with the `<style>` tag, you can override those classes with your own properties.

> **Tip**
> Using *CSS preprocessors* improves CSS development by using features such as variables, loops, and inheritance. However, Blazor doesn't have native support for CSS preprocessors such as *Sass* or *Less*, but this can be achieved through some third-party packages and extensions such as *Web Compiler* for Visual Studio. You can learn more about this at `https://github.com/madskristensen/WebCompiler`.

Now we have learned how we can add and manage the CSS of our project and examined the four different techniques available. Much of the time throughout this book, we will be working with isolated styles and inline styles, so this will be very useful.

Summary

Throughout this chapter, we have discovered various aspects of Blazor components, including how to create them using Razor syntax. We have also written our first component and learned how we can pass data between our components from parents to children and vice versa using parameters, cascading parameters, and EventCallbacks.

We also understood the life cycle events of each Blazor component, which are helpful to execute some logic in certain stages in the component's lifetime. Finally, we walked through the available ways to write our project's CSS, and saw how Blazor handles things behind the scenes for isolated CSS.

By now, you should be able to do the following:

- Create and use a Blazor component

- Pass data between components and across the components hierarchy

- Leverage the component life cycle events and react to certain changes within the component lifetime in the app

- Style your components and learn the different ways to deal with CSS in your Blazor app

In the next chapter, we will dive deeper into the concept of components, examining some special types such as layout components and templated components. We wil also learn how we render components dynamically in the UI, and consider the data binding mechanism for efficient two-way data communication between components.

Further reading

- Cascading values and parameters: `https://docs.microsoft.com/en-us/aspnet/core/blazor/components/cascading-values-and-parameters?view=aspnetcore-7.0`

- Component life cycle: `https://docs.microsoft.com/en-us/aspnet/core/blazor/components/lifecycle?view=aspnetcore-7.0`

- Isolated CSS styles: `https://docs.microsoft.com/en-us/aspnet/core/blazor/components/css-isolation?view=aspnetcore-7.0`

3
Developing Advanced Components in Blazor

In this chapter, we are going to dive deeper into the concept of components by learning about **layouts**, which define the general shape and the common components of the app pages and how to pass **user interfaces** (**UIs**) and other components as parameters using templated components. We will apply these concepts by creating our first layout and templated component. We will also see how we can render components in the UI dynamically using the **Dynamic Component** technique and learn how to create the **Razor Class Library** project and package or reuse our components in different projects.

This chapter will provide you with the required knowledge to develop more advanced components, in addition to grouping them in a Razor Class Library for future reusability.

In this chapter, we will cover the following topics:

- Building layouts in Blazor
- Developing templated components
- Rendering components dynamically
- Using the Razor Class Library for packaging and reusability

Technical requirements

The code used throughout this chapter is available in the book's GitHub repository here:

```
https://github.com/PacktPublishing/Mastering-Blazor-WebAssembly/
tree/main/Chapter_03.
```

Building layouts in Blazor

Most applications have common pieces of UI shared across different pages such as menus, navbars, company branding, and some other content. Rewriting this markup in every page component is very inefficient in modern software development. The solution is to have a single place where we place this shared content and then reference it from the pages, and here is where layout comes into play.

So basically, a layout is a special type of Blazor component that defines the general structure of the app by hosting the shared markup.

Understanding our Blazor project's default MainLayout

Before getting started with creating our own layout component, let's understand the default layout that comes with a Blazor project. The `MainLayout.razor` file within the `Shared` folder is the default layout for every new Blazor project, so if we open that file, we are going to notice the following:

```
@inherits LayoutComponentBase

<div class="page">
    <div class="sidebar">
        <NavMenu />
    </div>

    <main>
        <div class="top-row px-4">
            <a href=https://docs.microsoft.com/aspnet/
                target="_blank">About</a>
        </div>

        <article class="content px-4">
            @Body
        </article>
    </main>
</div>
```

As you may notice in the preceding code, the first special thing about the layout components is that they have the `LayoutComponentBase` base type, which contains @Body – this renders the routable components of the app (Index, Counter, FetchData, and so on).

The default `MainLayout` defines the NavMenu component with the `<sidebar>` element, a top bar with an **about** button, and an `<article>` element, which contains the main content of the app.

Creating our first layout component

In our **BooksStore** application, we need to have two different layouts: the default one will be used for the admin pages, and a new one that is simpler and cleaner will serve the users, as shown in the following screenshot:

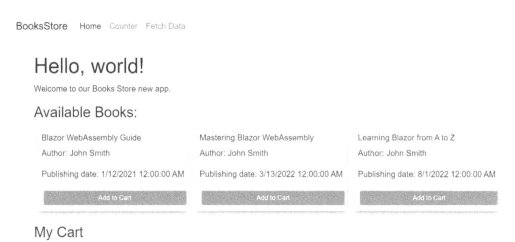

Figure 3.1 – The Index page with a new layout

So, let's create the new `UserLayout` and use it in our components in the following steps:

1. Before we create the layout itself, we need to create the new `NavBar` component at the top. So in the `Shared` folder, create a new component called `NavBar` and use the following code for it:

```
<nav class="navbar navbar-expand-lg navbar-light bg-light">
    <div class="container-fluid">
        <a class="navbar-brand" href="">BooksStore</a>
        <button class="navbar-toggler" type="button"
          data-bs-toggle="collapse"
          data-bs-target="#navbarNav"
          aria-controls="navbarNav"
          aria-expanded="false"
          aria-label="Toggle navigation">
            <span class="navbar-toggler-icon"></span>
        </button>
        <div class="collapse navbar-collapse"
          id="navbarNav">
            <ul class="navbar-nav">
                <li class="nav-item">
                    <a class="nav-link active"
```

```
                              aria-current="page"
                              href="">Home</a>
                    </li>
                    <li class="nav-item">
                        <a class="nav-link"
                           href="/Counter">Counter</a>
                    </li>
                    <li class="nav-item">
                        <a class="nav-link"
                           href="/FetchData">Fetch Data
                        </a>
                    </li>
                </ul>
            </div>
        </div>
    </nav>
```

2. Create a new Razor component in the `Shared` folder as well and give it the name `UserLayout` (you can place the component in another folder as well, but the `Shared` folder is convenient for housing shared components in our app).

3. Set the base class of `UserLayout` by inheriting from `LayoutComponentBase`, as every layout should have this as its base type:

    ```
    @inherits LayoutComponentBase
    ```

4. Render the new `NavBar` component at the top of the layout:

    ```
    ...
    <NavBar />
    ```

5. Create a `<main>` element and give it some padding and render the `@Body` property within it:

    ```
    ...
    <main class="px-5 py-4">
        @Body
    </main>
    ```

    So, the full code of the new simple layout should look as follows:

    ```
    @inherits LayoutComponentBase

    <NavBar />

    <main class="px-5 py-4">
        @Body
    </main>
    ```

Now, we are ready to start using this Layout in our components.

Defining the layout for a component

Blazor provides us with different ways to set the layout of our components, including applying the layout to a certain component, a group of components, or all the components of the app:

- *Applying the default layout*: The default layout of the project is defined within the App.razor file, as you can see in the following code:

```
...
<Found Context="routeData">
        <RouteView RouteData="@routeData"
           DefaultLayout="@typeof(MainLayout)" />
        ...
```

The RouteView component has the DefaultLayout parameter, which we can use to set the type of the default layout that will be used for all application components.

- *Applying a layout to a folder of components*: We can define the layout of all the components existing in a certain folder. As we mentioned in *Chapter 1, Understanding the Anatomy of a Blazor WebAssembly Project*, we used the _Imports.razor file to define shared used namespaces for all the components. The _Imports.razor file can be created in each folder or in a specific one, and we can use it to define the layout for all the components within that folder using the @layout directive. In our example, we will create a new folder inside the Pages folder and call it UserPages. We will place the pages that will be accessed by the public users and not the admins within it, and then we will move Index.razor inside the UserPages folder as well:

Figure 3.2 – The Index.razor component inside the UserPages folder

Now, to set the layout of the Index component and all future components within this folder, we will create an _Imports.razor file in the UserPages folder and then set the layout to the UserLayout that we created in the previous section as follows:

```
@layout UserLayout
```

Now, if we run the application, you are going to see the **Index** page with the same layout as in *Figure 3.1*. However, if you navigate to /Count, you will notice the **Counter** page within the default Blazor layout with the purple nav menu on the left.

- *Applying a layout to a single component*: In some cases, only a certain page or pages need different components, but they are not grouped in a folder. Imagine a scenario where an application has the Login/Register pages in a blank layout that doesn't have a nav menu. In such a scenario, you can set the Login page layout specifically using the same @layout directive but at the level of the component.

In the following example, we are going to change the layout of the FetchData page in the FetchData.razor file, and set it to use UserLayout but at the level of the component:

```
@page "/fetchdata"
@inject HttpClient Http
@inject ILoggingService LoggingService

@layout UserLayout

<PageTitle>Weather forecast</PageTitle>
...
```

After running the application and navigating to /fetch-data, you will notice that the FetchData component is wrapped with the new layout created for public users:

BooksStore   Home   Counter   Fetch Data

Weather forecast

This component demonstrates fetching data from the server.

Date	Temp. (C)	Temp. (F)	Summary
5/6/2018	1	33	Freezing
5/7/2018	14	57	Bracing
5/8/2018	-13	9	Freezing
5/9/2018	-16	4	Balmy
5/10/2018	-2	29	Chilly

Figure 3.3 – The FetchData component using UserLayout

- *Applying a layout to a portion of a component*: Blazor provides us with a special component called LayoutView that sets the layout for its children only and can be used inside any component.

The default usage of LayoutView is in the Router component within App.razor. Basically, if we try to navigate to an undefined URL, the Blazor app will render the NotFound component section in App.razor, and by default, Blazor will set the layout of the NotFound content using the LayoutView component. The following code shows the usage of LayoutView to set MainLayout to the application's NotFound content:

```
<Router AppAssembly="@typeof(App).Assembly">
    ...
    <NotFound>
        ...
        <LayoutView Layout="@typeof(MainLayout)">
            <p role="alert">
             Sorry, there's nothing at this address.
            </p>
        </LayoutView>
    </NotFound>
</Router>
```

> **Nested layouts**
>
> Layouts can be nested within each other by defining the layout of a layout the same way we define the layout of a component.
>
> An example of using nested layouts could be an application having nested menus. In an application with many sections, each section would have its own menu, and the app as a whole would have a menu as well. In such a case, we can actually define a default layout for the app, and a different layout for each section that references the main layout of the app.
>
> To learn more about this, read https://docs.microsoft.com/en-us/aspnet/core/blazor/components/layouts?view=aspnetcore-7.0#nested-layouts.

In this section, we learned about Blazor's default layout, creating new layouts, and applying layouts partially, to single pages, or to sets of components. Now, we can move to the next enjoyable section of Blazor, which is about templated components that are more flexible and customizable compared to other standard components.

Developing templated components

We have seen how we can create a component and pass some data to it so it renders them in a specific way; however, what if we want to pass a piece of the UI to that component and inject some markup into it? We call these types of components that take UI elements as parameters **templated components**.

In templated components, we can receive the UI content to be rendered in a normal parameter, but it's of the `RenderFragment` or `RenderFragment<Tvalue>` types. Each component can have one or multiple `RenderFragment` parameters.

In this section, we will learn about developing templated components practically by creating two different components that we will use for our project in this book, and which can be used in your other projects as well.

Developing a modal pop-up component

A modal popup is a popular component, as it is the topmost element we see in modern apps. The component allows the user to view further details (for example, for a product on sale), fill out a form, or take some actions quickly by showing a box in the middle of the screen and disables clicking anywhere else outside it. Such popups help the user stay on the same page without needing to navigate and keep the required task simple.

Our modal pop-up component will be templated because we will just develop the modal as a high-level component. The content of the modal, whatever it is (a form, details, a question, and so on), will be passed as a `RenderFragment` parameter. Basically, we can think of this component as a template we can reuse based on our needs. The final result of our example will look like the following:

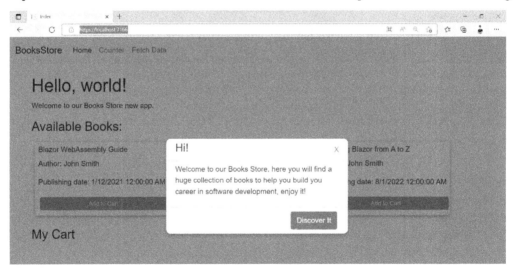

Figure 3.4 – ModalPopup templated component shows a welcome message

So, let's get started:

1. In the `Shared` folder, add a new Razor component and call it `ModalPopup.razor`.

2. For the CSS file associated with this component, we need to create it in a scoped style, as mentioned in *Chapter 2, Components in Blazor*. To do so, we must create a file named `ModalPopup.`

`razor.css`. After creating the Razor component and the CSS files, you should have the following result:

Figure 3.5 – Files structure of the ModalPopup within the Shared folder

3. In the `ModalPopup.razor` file, let's create five parameters: `Title` of a `string` type, `ChildContent` of a `RenderFragment` type, `FooterContent` of a `RenderFragment` type, and `IsOpen` of a `bool` type. `Title` will be used as a title for our modal, `ChildContent` will be the UI markup that we will pass when we use the modal, and `FooterContent` will comprise another UI markup to render the bottom part of our modal. With those parameters and with the avoidance of a fixed UI, the modal will be usable for every case needed, instead of needing to create a new modal for every possible usage.

    We will also need an `EventCallback` parameter called `OnCloseClicked` to use to notify the parent of the modal when the close button gets clicked, and finally, the `IsOpen` parameter to show or close our modal from the parent component by assigning the value of this parameter as true or false:

```
@code
{
    [Parameter]
    public string? Title { get; set; }
    [Parameter]
    public RenderFragment? ChildContent { get; set; }
    [Parameter]
    public RenderFragment? FooterContent { get; set; }
    [Parameter]
    public bool IsOpen { get; set; }
    [Parameter]
    public EventCallback OnCloseClicked { get; set; }
}
```

4. Now, let's design the UI of this modal; when the popup opens, there will be an overlay layer in the background and the modal box in the middle of the screen. The box will be divided into three parts: `header`, `body`, and `footer`. The `header` will host the title and a close button, the `body` will render the `ChildContent` parameter, and the `footer` will render the `FooterContent` parameter. Of course, all this UI will be rendered only if the modal is open, so we will have the following markup:

```
@if (IsOpen)
{
    <div class="overlay">
        <div class="modal-box">
            <div class="modal-header">
                <h4>@Title</h4>

                <span class="close-button">X</span>
            </div>
            <div class="modal-body">
                @ChildContent
            </div>
            <div class="modal-footer">
                @FooterContent
            </div>
        </div>
    </div>
}
@code
{
    ...
```

The `ChildContent` and `FooterContent` parameter values will comprise markup code, so the `ModalPopup.razor` component doesn't have a fixed look; every time we use it, we need to pass the `body` and the `footer`, and that's what makes it a templated component. Unlike `BookCard`, which we created in *Chapter 2*, and has a fixed shape and look and just received a book object to render it a certain way, our modal popup can be used anytime with different content. Its main function is to provide us with a pop-up capability that we need to wrap our content with.

5. Before we move to the styling part, we just need to activate the close button, so it closes the modal when the user clicks it and fires `OnCloseClicked` `EventCallback` to notify the component that holds the modal that it got closed from inside it.

To achieve the desired result, we just need to write a small function that sets the value of IsOpen to false, as shown in the following code:

```
...
    private void CloseModal()
    {
        IsOpen = false;
        OnCloseClicked.InvokeAsync();
    }
```

And in the span element, we need to set the value of the @onclick event to the CloseModal method:

```
<span class="close-button" @onclick="CloseModal">X</span>
```

6. By writing the previous markup, we just need to write the required styles for the classes mentioned in the HTML code. So, back in the ModalPopup.razor.css file, let's add the following class for the overlay layer in the background, which is basically what gives the .overlay div a fixed position, covers the full page, and gives it a black background with 50% opacity:

```
.overlay {
    position: fixed; /* Set its position based on the
                        screen */
    width: 100%; /* Set the width to the full screen
                    width */
    height: 100%; /* Set the height to the full screen
                     height */
    top: 0; /* Set its top starting point to 0 */
    left: 0; /* Set its left starting point to 0 */
    background-color: rgba(0,0,0,0.5); /* Black
    background with opacity */
    display: flex; /* Define the overlay as flex
    container to host the modal box */
    justify-content: center; /* Set the children of
    the overlay horizontally in the center */
    align-items: center; /* Set the children of the
    overlay vertically in the center */
}
```

7. Add the styles of the .modal-box div, which will show in the center because of the flex container of its parent (the .overlay div). We will create a white background, a little shadow, and some smoothness on the borders as follows:

```
...
.modal-box {
    width: 400px; /* Give it a width, this can be set
                     as parameter */
```

```
        background-color: white; /* Give it a white
                                  background */
        border-radius: 10px; /* Set the radius of the
                              corners */
        box-shadow: 0 3px 7px rgba(0, 0, 0, 0.3);
        /* Little shadow */
    }
```

8. After defining the style of the full box, it's time to style the parts of its content, and we will start with the header. The goal is to set the title on the very left and the X button on the very right. We can also achieve this using flex and setting the justify-content property to space-between as follows:

```
    ...
    .modal-box .modal-header {
        display: flex; /* Set the header as a flex
                        container */
        align-items: center; /* Align the title and
        the close button vertically in the center */
        justify-content: space-between; /* Set the
        title and the close button on each side of the
        header */
        padding: 5px 20px; /* Set a little padding to
                            look a bit better */

    }
```

9. Within the header, we need to style the X button a little bit by setting the color of X to red and setting Set the cursor to pointer, so it gives feedback that this is a clickable element:

```
    ...
    .modal-box .modal-header .close-button {
        color: red; /* Set the color of the font to
                    red */
        cursor: pointer; /* Set the cursor to pointer
        to indicate that this is clickable */
    }
```

10. Finally, set the styles of the body and the footer content of the modal. We only need padding; the rest of the content will be defined when we set its body using the modal in another component:

```
    ...
    .modal-box .modal-body, .modal-footer {
        padding: 5px 20px;
    }
```

Our modal component is now ready to be used to render special content in a modal popup. Now, we will consume it and see the power of the templated components by using the same components with different content each time.

Consuming the modal pop-up component

Our modal component is currently serving the purpose of a modal box, which shows in the middle of the screen and can be populated with anything we want.

To keep things simple, we are going to create a little welcome message that will be shown to our users every time they open the Index page, with a button at the bottom of the dialog to close it.

When using a templated component, the values of the RenderFragment parameters can be assigned by opening two tags with the name of the parameter within the component tag and then typing the required markup inside it. The next example will show you how to populate the ChildContent parameter and the FooterContent parameter for the dialog that we have created:

1. In the Index.razor file, we will render the ModalPopup component and set a P tag in the body with a welcome message and a button in the footer to close it. The following code snippet could be added to the end of the Razor code on the **Index** page:

    ```
    ...
    <ModalPopup Title="Hi!">
        <ChildContent>
            <p>Welcome to our Books Store, here you will
               find a huge collection of books to help you
               build you career in software development,
               enjoy it!</p>
        </ChildContent>
        <FooterContent>
            <button class="btn btn-primary">Discover It
            </button>
        </FooterContent>
    </ModalPopup>
    ```

    As you may notice, we supplied the Title parameter's value as an attribute, and for the value of the ChildContent and FooterContent parameters, we had to open two tags and set the elements inside. This is what makes modal components very interesting, as with them, you can inject any UI element or other components inside, and they will be rendered within this component (ModalPopup in our example).

2. Next, we need a variable to control the `IsOpen` parameter, which is responsible for opening and closing our modal popup. Basically, we are going to create a simple variable of a `bool` type and assign it to `true`, as we want the modal to be open by default. So, in the `@code` section, let us declare the variable and call it `_isWelcomeModalOpen`:

    ```
    ...
    @code
    {
        private bool _isWelcomeModalOpen = true;
        ...
    ```

3. Back in our modal component, we need to assign the value of the `_isWelcomeModalOpen` variable to the `IsOpen` parameter of the `ModalPopup` component. So, whenever we change the value of the variable, the status of the modal will also be affected:

    ```
    ...
    <ModalPopup Title="Hi!" IsOpen="@_isWelcomeModalOpen">
        <ChildContent>
            ...
    ```

4. There are two ways to close this modal: one through the **X** button built-in within `ModalPopup` itself, and the other is through the **Discover it** button, wich we have in the footer of the modal (see *Figure 3.4*).

    When the user clicks the **X** button from inside the modal, the popup closes but keeps the value of our variable set to `true`, so any change in the state of the component that requires it to re-render itself will cause the modal to open again. To resolve this, we need to set the value of `_isWelcomeModalOpen` to `false` every time the user closes the modal using the **X** button, and that could happen by subscribing to the `OnCloseClicked` `Eventcallback` of `ModalPopup`.

    We are going to create an anonymous method using *Lambda* in *C#* to set the value of the variable whenever `OnCloseClicked` gets fired as follows:

    ```
    ...
    <ModalPopup Title="Hi!" IsOpen="@_isWelcomeModalOpen"
    OnCloseClicked="() => _isWelcomeModalOpen = false">
    ...
    ```

5. The last step in our example is also to set `_isWelcomeModalOpen` to `false` when the user clicks the **Discover it** button, so all we have to do is to repeat the little method but assign it to the `onclick` event of the button, as shown in the following code snippet:

    ```
    ...
        <FooterContent>
            <button class="btn btn-primary"
                @onclick="() => _isWelcomeModalOpen =
    ```

```
        false"> Discover It</button>
    </FooterContent>
    ...
```

Here you go! You have created and used your first templated component! After running the project and navigating to the Index page, you will notice the result shown in *Figure 3.4.*

> **Note**
>
> If you have a templated component that accepts a `RenderFragment` parameter, you can set its value directly inside the component without the need to open the tag with the name of the parameter, but the parameter must be called `ChildContent`.
>
> In our `ModalPopup` example, if we directly set elements inside the `ModalPopup` tag without using `<ChildContent>` or `<FooterContent>`, it will consider the content as the value for the `ChildContent` parameter so that markup will be rendered in the body of the modal.

Generic templated components

Until now, we have seen how we can pass UI markup to a Blazor component, but the concept of templated components goes beyond this to include the `RenderFragment<TType>` parameter (you will discover the benefits of the `TType` parameter in the upcoming example). The concept of generic is very popular in the world of programming and C#. Specifically, it allows the developer to build a class that targets different types with the same code. For example, when we declare an instance of the famous `List<TType>` in C#, we define its generic type as `List<Book>`, `List<int>`, and so on. Resultantly, all the members of the class (methods, properties, and so on) that target the `T` type will target the type defined in the initialization such as string, int, the custom types we create, and so on.

With generic, .NET has to introduce one `List` class that is generic, so developers can use it to store any type they want. All the methods of the `List` class, such as `Add` and `Remove`, will work against the type we define. Without generic, .NET has to write the `List` class for each type, such as `ListOfString`, `ListOfObject`, and `ListOfInt`, which will lead to long repetitive code and won't be able to cover the custom types that we create in our apps such as the `Book` class in our project.

Blazor gives us the ability to develop generic templated components, such as `List`, defined in the .NET ecosystem. What does this mean? Let's go back to our **BookCard** component, which we created in *Chapter 2* to render a book as a card and then use it in other UIs. The component was basically accepting `Book` as a parameter and rendering it the same way every time except when the data was being passed to it.

What if we want to develop a component called **DataListView**, and this component accepts a collection of the T data types (the user can define the type of data that this component renders while using it) and also allows us to define the template of how to render a single item of this collection? After that, **DataListView** renders that list for us either using a data grid concept (in rows and columns such as *Instagram* content in your profile) or as a vertical list (such as in your email inbox in *Microsoft Outlook*). In addition to rendering the list, the component will provide us with the option of selecting an item, selecting multiple items, and some other functionality that we use whenever we have a list of data to deal with in the UI.

This will make things much easier right? Having a single component to handle all of that for us will save us time and energy to repeat the same functionality for each type of data. All we need to do is declare the component, pass the list of data to it, and tell it how to render one item of it, and done.

Let's get started by developing the basic version of this component, which for now, will render the items in a grid view (rows and columns), and then we can extend its functionality whenever needed:

1. In the `Shared` folder, create a new Razor component and call it `DataListView.razor`.

2. Now we need to define this component as a generic component, so we build it around a generic type, and when we want to use it, we pass the type of that parameter to, for example, `Book`, `Car`, `int`, and so on.

    To achieve this, we need to use the `@typeparam` Razor directive at the top of the component as follows:

    ```
    @typeparam TItemType
    @code
    {
    }
    ```

3. Define the parameters of our component. Basically, we will have the following parameters: `Items` of the `List<TItemType>` type, which is the collected data the developer wants to render, `ItemTemplate` of the `RenderFragment<TItemType>` type, which is the UI of each item in the list, `EventCallback<TItemType>` called `OnItemClicked`, which will be fired every time the user clicks on any item from the collection, and finally, a `ColumnsCount` of the `int` type, which will be used to show how many items should be rendered horizontally in the grid:

    ```
    . . .
    @code
    {
        [Parameter]
        public List<TItemType>? Items { get; set; }
        [Parameter]
        public RenderFragment<TItemType>
          ItemTemplate { get; set; }
        [Parameter]
        public EventCallbackTItemType> OnItemClicked {
          get; set; }
        [Parameter]
        public int ColumnsCount { get; set; } = 4;
    }
    ```

4. Now, it's time to design the markup. To render the items in a grid fashion, we are going to use the capabilities of the CSS flexbox to align the items horizontally and wrap them into new rows when the rows get completed. We will render each item within a `div` with the width based on the `ColumnsCount` parameter value (25% width if, for example, the columns count is 4).

```
...
<div class="grid-container">
    @if (Items != null && Items.Any())
    {
        foreach (var item in Items)
        {
            <div class="grid-item">
                @ItemTemplate(item)
            </div>
        }
    }
    else
    {
        <!-- Here can be used to show how the empty
            list will be handeled -->
        <h3 class="empty-collection-title">
          No items available</h3>
    }
</div>
...
```

Inside the div, we first checked whether the collection of items is null or empty showing the user a message indicating that there are no elements in the list rather than showing them an empty UI. This action improves the user experience. In an advanced scenario, you can pass a `RenderFragement` parameter called `EmptyContentTemplate` and render it instead of the H3 tag we must give the developer when using this component. This gives us the ability to control the content of the empty list.

Another thing to notice in the *eighth* line of the preceding code snippet is that we rendered `ItemTemplate`, but we passed a parameter to it, which is the difference between a normal templated component and a generic one. When the developer defines the template for each item in the collection, that item will be available as a parameter so they can use it to render the data inside that item.

5. Now, it's time to add some styling to our component. Again, we will apply isolated styling by creating a new file in the `Shared` folder called `DataListView.razor.css`.

6. Create a CSS class called `.grid-container`, which will define some properties to the parent `div` in the markup.

```
.grid-container {
    display: flex; /* Enable the flex box for the
```

```
                          elements inside the div */
        flex-wrap: wrap; /* Wrap the items inside when
                          they reach the end of the row */
        padding: 5px; /* Define a little padding for the
                          list */
        border: 0.5px solid #A3A3A3; /* Set a slight
                                       border for it */

    }
```

7. Create a `.grid-item` CSS class to define the style of each item in the grid; we are not going to set the width property for each item because it will be generated based on the value passed by the developer, so we will set it inline in the markup:

```
    . . .
        .grid-container .grid-item {
            margin: 4px; /* Define a spacing around each
                            item */
            cursor: pointer; /* Use pointer cursor to
              indicate that the item is clickable */
        }
```

8. The last style we will define is the style of the H3 tag for the empty list, which is just to centralize the text and set a little margin from the top:

```
    . . .
        .grid-container .empty-collection-title {
            text-align: center;
            margin-top: 10px;
        }
```

9. Back to the `DataListView.razor` file, we now need to calculate the width of each item based on the `ColumnsCount` passed to the component.

10. To accomplish this, we just need to divide `100` by the number of columns, and we will get the width percentage required for each column and then assign the calculated percentage as a width style to `div`.

    So, after defining the parameters, we will create a read-only property called `_columnWidth`, which is equal to `100 / ColumnsCount`:

```
    . . .
        private int _columnWidth => 100 / ColumnsCount;
    . . .
```

Then, set the style attribute of the `div`, which wraps the item template as follows:

```
...
<div class="grid-item" style="width:@_columnWidth%">
    @ItemTemplate(item)
</div>
...
```

After rendering the preceding code snippet, if the developer passes 5 to the `ColumnsCount` parameter, then the value of the style attribute will have `width:20%`.

11. Finally, we just need to add the clickable functionality to the `div` grid item, so that whatever the content is, `div` can be clicked to fire `OnItemClicked EventCallback`. All we need to do is to assign the `@onclick` event of the `div` grid item to an anonymous function, which triggers `EventCallback` and pass the clicked item to it as a parameter:

```
...
foreach (var item in Items)
{
    <div class="grid-item"
      style="width:@_columnWidth%" @onclick="() =>
      OnItemClicked.InvokeAsync(item)">
        @ItemTemplate(item)
}
...
```

That's all it takes to create an advanced generic templated component. Keep in mind when you are developing this kind of component to provide functionality and rendering logic for your different UIs and to minimize the number of changes needed while using it in different cases. You need to parametrize its logic as much as you can. For example, in `DataListView`, we can have parameters for the spacing (margin and padding), one for the border, one Boolean parameter to allow selection or not, and another Boolean to make the item clickable or not.

Further, in real-world scenarios, the component should handle the values of its parameters correctly, so you can use the `OnParametersSet` method to validate the value of the parameters.

> **Note**
>
> The advanced code for the `DataListView` component can be found in the `Complete` folder within the repository of this book on GitHub and contains all the mentioned features: `https://github.com/PacktPublishing/Mastering-Blazor-WebAssembly/Complete`.

Consuming a generic templated component

An index component renders a list of books using the `BookCard` component. Now we will leverage our `DataListView` component to render that list of books for us. All we have to do is just pass the list of the books and define the template of each book, and it will handle the rest.

Let's go through this task to develop better functionality and code for our component:

1. In the `Index.razor` file, on top of the **My cart** section, let's add a `DataListView` component and assign the value of `TItemType` to `Book` so the generic component defines the type it will use internally. Assign the `_books` list we have in the Index to the `Items` parameter of the `DataListView` component and set `ColumnsCount` to 4:

    ```
    ...
    <DataListView TItemType="Book" Items="_books" ColumnsCount="4">

    </DataListView>
    ...
    ```

2. Now, we need to define the template of each item. We can do this by opening a tag with the `ItemTemplate` name, which will be the piece of UI that will be rendered inside the `DataListView` for each book.

3. Basically, we need to keep rendering each book as a `BookCard` component, but this time, we don't have the variable that exists in `foreach` to access each `book` object. Instead, we have a `context` variable within the tags of `ItemTemplate`, which represents the book, and the value of this variable is the value that we passed when we rendered the template in the 4th step of creating the component in the previous section:

    ```
    ...
    <DataListView ItemType="Book" Items="_books" ColumnsCount="4">
        <ItemTemplate>
            <BookCard Book="@context"
                OnAddToCartClicked="AddToCart" />
        </ItemTemplate>
    </DataListView>
    ...
    ```

In some cases, you may have another `context` variable in your code, which will make the compiler unable to know which `context` variable to use. In this case, you can rename the default context variable explicitly using the `Context` attribute of the `ItemTemplate` tag as follows:

```
...
<ItemTemplate Context="book">
    ...
```

4. The last step is removing the old iteration loop that we created below the H3 header of the My Books title.

    After running the project, the result won't be very different, but what we have now is a more efficient way to render the list: it's a reusable component that we can use in many places in our project, and of course, we have more functionality to leverage and centralize in one place. So if we want to edit any of the list view functionality, we can do it in its component file, and it will affect all the places it was used in. Most importantly, you have learned how to think through and develop a real-world templated component from scratch:

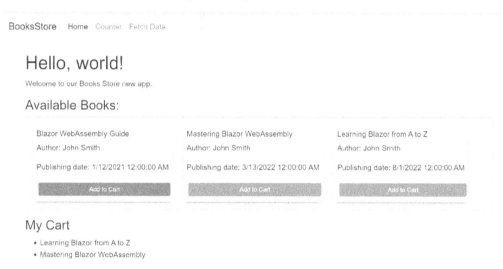

Figure 3.6 – A list of books using the DataListView templated component

Until now, we have learned about some cool concepts in Blazor that allow you to build some interactive and very reusable parts of a full single-page application, but that's not everything. Blazor offers much more. In the next section, we will look at another interesting component type called a dynamic component, which helps us render some UIs dynamically.

Rendering components dynamically

This section's heading seems interesting, but haven't we already seen how to render dynamic stuff in the UI? Well, what's different here is that the component itself is not dynamic, but its rendering is.

Let's take a real-world scenario that we usually see in many advanced and big apps (*Azure Portal*, for example), which have a dashboard that is customizable by the user. This app offers a huge collection of widgets (each widget is a component) and the user can create their own dashboard by choosing which widgets to place on the dashboard page.

There are many solutions for such a scenario. For example, we can store the selected widget names in an array and then iterate over the array in the dashboard page, then make an `if` statement based on the name of the widget, and render the corresponding component. Actually, this works, but if you think about it, the need for a customizable dashboard means a massive number of widgets. Otherwise, there can be a fixed dashboard, and all the widgets are allocated somewhere on it.

The `if` statement will make the code hard to maintain and very long, and we may make a mistake by rendering the wrong component in a certain `if` clause.

This is where the dynamically rendered component comes into play; basically, it's a built-in component called `DynamicComponent` provided by Blazor in *.NET 7.0* that renders a specific component by passing its type as a parameter to `DynamicComponent`. Of course, we can pass parameters too.

So, back to our dashboard example, we can store the name of each component type in an array and iterate over it on the dashboard page. Then, in each iteration, all we need to do is to render `DynamicComponent` and pass the type of the widget component to it, so the code will be a single component rendered within a `foreach` loop; internally, `DynamicComponent` will take care of the rest.

Let's see a quick example of how to use `DynamicComponent` to render the `SurveyPrompt` component initially created by the default Blazor template using its type:

1. In `Counter.razor` in the `Pages` folder, just to keep the Index page clean for now, let's define a variable of type `Type` in C#, which represents the variable's data type (for more information about the `Type` class, please visit `https://docs.microsoft.com/en-us/dotnet/api/system.type?view=net-7.0`).

    Then, assign the `typeof(SuveryPrombt)` value to it. The `typeof()` operator in C# retrieves the `Type` parameter of a C# data type as the following code snippet shows:

    ```
    . . .
        private Type componentType = typeof(SurveyPrompt);
    ```

    The `componentType` variable can hold the type of any component we have in our project, so using this variable to render a component is what makes things very dynamic in that fashion.

2. In the Razor markup, create `DynamicComponent` and pass the `componetType` variable to its `Type` parameter as follows:

    ```
    . . .
    <DynamicComponent Type="componentType" />
    . . .
    ```

    Now, if we run the project and navigate to the `/counter` page, you will notice the `SurveyPrompt` component rendered in the UI, and that magic happens just using its own type. So, if we change the value of the `componentType` parameter at runtime, we will notice another component will be rendered instead of `SurveyPrompt`, and that's what gives `DynamicComponent` the power to handle such scenarios:

Figure 3.7 – The SurveyPrompt component rendered using DynamicComponent

This also leads to a question: what if the rendered component takes parameters? DynamicComponent handles this case by providing us with a Dictionary<string, object> parameter, which contains a list of key values (key of type string represents the name of the parameter, and the value is of type object that could be the value of that parameter), so we can use that dictionary to supply the parameters of the component being rendered by DynamicComponent.

The following code snippet supplies the Title parameter of type string that the SurveyPrompt component accepts using the Parameters of DynamicComponent:

```
...
<DynamicComponent Type="componentType" Parameters="parameters" />
@code {
...
    private Type componentType = typeof(SurveyPrompt);
    private Dictionary<string, object> parameters = new()
    {
        { "Title", "Welcome to Blazor" }
    };
}
```

Many advanced features can be delivered using DynamicComponent, such as the dashboard example we provided, or allowing the admin to change how the sections on the home page of their app look without the need to change the code. All these tasks are covered within the Complete folder in the GitHub repo of this book.

Now let's make our code even more reusable by discovering the Razor Class Library project.

Using the Razor Class Library for packaging and reusability

We have created plenty of components in this chapter and the previous one, but as you may have noticed, all those components are mostly located in the same project within the Shared folder. So, you may have wondered how to use the exact same component in other projects within the same solution or maybe within a totally different solution.

.NET offers another type of project called the Razor Class Library, and the main purpose of this project is to have a Blazor component, and then you can reference this project within your other projects, solutions, your entire organization, or even publish it as a **NuGet** package so it could be used by other developers worldwide.

In this section, we will create a new Razor Class Library project, reference it within our *Books Store* project, and learn how to reference static resources such as CSS and JavaScript files within our Blazor WebAssembly project.

Sometimes, reusability is not limited to the level of certain components; it becomes mandatory in some cases, such as when developing web and mobile clients with Blazor and .NET **Multi-platform App UI (MAUI)**. In such cases, you will have a Blazor WebAssembly project and a .NET MAUI one. So, a Razor Class Library project is needed here to host the components, and the pages are shared between the mobile and the web app.

Creating a Razor Class Library project

Inside our **BooksStore** Visual Studio solution, we will add a Razor Class Library project and give it the name BooksStore.Blazor.Components. To achieve this, take the following steps:

1. Right-click on the solution and click on **Add | New Project**.
2. In the pop-up box for adding a new project in Visual Studio, search for Razor Class Library and choose the project as shown in the following screenshot:

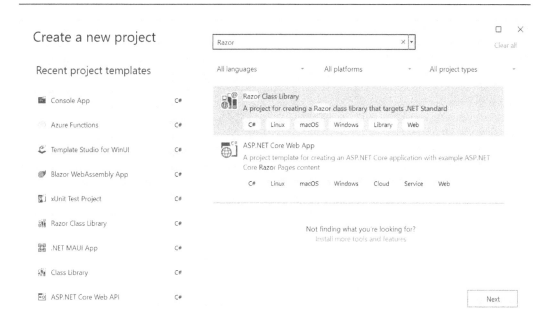

Figure 3.8 – Add the Razor Class Library project

3. Give it the name BooksStore.Blazor.Components and then click **Next**, then **Create**.

By default, the project gets created and initialized with the following:

- wwwroot folder: This is the folder where we can have static resources such as the global CSS file(s), images, and JavaScript files.

- By default, it has the Background.png and exampleJsInterop.js files as examples, and we can remove them safely for now.

- _Imports.razor: This is the file where we can host the shared parts that can be used in other components or define the shared layout for all the existing components within the project.

- Component1.razor: This is a simple example component; we can remove it if we like.

- ExampleJsInterop.cs: This is a C# service to demonstrate the integration with JavaScript. We can remove this, too, as we will learn more about Blazor and JavaScript in *Chapter 6, Consuming JavaScript in Blazor.*

After creating the project, we can now start creating our shared components inside it, but before doing so, let's reference this library project in our **BooksStore** project.

Referencing the Razor Class Library in the Blazor WebAssembly project

Referencing the Razor Class Library in a Blazor WebAssembly project requires two steps. The first is adding the assembly to the references of the Blazor WebAssembly project, and the second is adding references to the styles and enabling the scoped styles for the CSS isolated for each component and reference JavaScript files if any.

Let's accomplish the following steps to achieve our goal:

1. Right-click on **Dependencies** for the **BooksStore** project and click **Add Project Reference**

2. All other projects existing within the solution will be listed; choose **BooksStore.Blazor. Components** and then click **OK**.

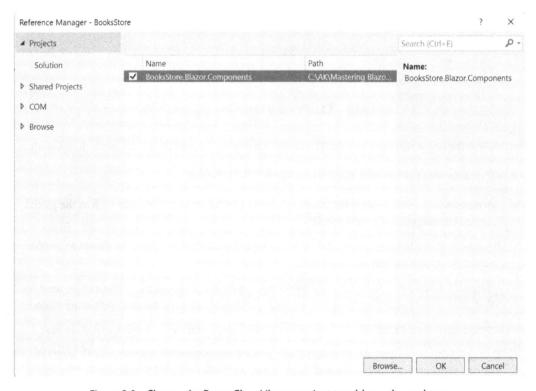

Figure 3.9 – Choose the Razor Class Library project to add as a dependency

3. Now, we need to enable the scoped styles that we will add to our components in the Razor Class Library, and to do so, we need to add a link tag in the header of the index.html file of the Blazor project existing in the wwwroot folder. And that link will refer to the following location _Content/{ProjectName}/Styles.css. So, in our example, it will be _Content/BooksStore.Blazor.Components/Styles.css as follows:

    ```
    <head>
    ...
    <link href="_Content/BooksStore.Blazor.Components/styles.css"
    rel="stylesheet" />
    </head>
    ```

    Also, we can add the namespace reference existing in our Razor Class Library project, so all the components and pages in Blazor WebAssembly have access to it without the need to add them every time in each component by adding them to the _Imports.razor file. We will skip this step for now, as our Razor Class Library project has no components yet.

4. If any CSS file exists within the wwwroot folder of the Razor Class Library project, we can reference it using the same previous step but by using the name of the file. For example, if we have a CSS file called components.css in the Razor Class Library project, we can reference it as follows:

    ```
    ...
    <link href="_Content/BooksStore.Blazor.Components/components.
    css" rel="stylesheet" />
    ...
    ```

5. Finally, if there are any JavaScript files, we can reference them using the same reference link of the CSS files, such as _Content/ProjectName/JavaScriptFileName.js, and we can add it to the bottom of the body tag in the index.html file. For example, if we have a file called component1-logic.js, we can reference it as follows:

    ```
    ...
        <script src=
        "_Content/BooksStore.Blazor.Components/
        component1-logic.js"></script>
    </body>
    </html>
    ```

The BooksStore.Blazor.Components Razor Class Library project will be used heavily throughout the rest of the book by adding most of the new components that we will create in the later chapters of this book.

> **Note**
>
> If your project is a Razor Class Library project that any developer or any team within your organization can use, you can publish it to NuGet to make it accessible to them. The following link shows you the details of doing so: `https://docs.microsoft.com/en-us/nuget/quickstart/create-and-publish-a-package-using-visual-studio?tabs=netcore-cli`.

Now, we have a clean project we can use and reference anywhere, which will host all of our reusable components. This step will help keep our Blazor WebAssembly small and clean and provide us with the ability to have all the reusable components just a project reference away.

Summary

In this chapter, we covered some interesting topics and concepts in Blazor, from understanding layouts and creating our first one to discovering the amazing concept of templated components while developing and using two interesting components. After that, we went a bit more advanced into showing the power of the `DynamicComponent` available in Blazor, and finally, we covered the Razor Class Library, understood its benefits, and created and referenced our own component.

After completing this chapter, you should now be able to do the following:

- Create and consume different layouts for your application

- Create templated components to help you wrap and reduce the amount of repeated code by defining a template for a certain task such as a modal popup or a data list view

- Take advantage of the ability of `DynamicComponent` to have more dynamic UIs in your apps

- Create and reference Razor Class Library projects to have a more organized project structure and more reusability for your components

In the next chapter, we will move on to the concept of navigation, and learn how we define pages and the parameters that they accept from the URL, as well as move and navigate between our app pages using the powerful built-in service, `NavigationManager`.

Further reading

- Layouts in Blazor: `https://docs.microsoft.com/en-us/aspnet/core/blazor/components/layouts?view=aspnetcore-7.0`

- Templated components: `https://docs.microsoft.com/en-us/aspnet/core/blazor/components/templated-components?view=aspnetcore-7.0`

- Dynamic components: `https://docs.microsoft.com/en-us/aspnet/core/blazor/components/dynamiccomponent?view=aspnetcore-7.0`

- Razor Class Library projects: `https://docs.microsoft.com/en-us/aspnet/core/blazor/components/class-libraries?view=aspnetcore-7.0&tabs=visual-studio`

Part 2:
App Parts and Features

In this part, you will discover the parts and features for developing a complete SPA in Blazor WebAssembly. You will learn about navigation, how to build and validate forms, how to develop custom input components, and how to utilize JavaScript in your Blazor apps. In addition to this, you will learn about state management and how to consume REST APIs from your Blazor client-side app, and then we will deep-dive into authentication and authorization in Blazor. Finally, you will learn how to reliable apps by performing efficient error handling.

This part has the following chapters:

- *Chapter 4, Navigation and Routing*
- *Chapter 5, Capturing User Input with Forms and Validation*
- *Chapter 6, Consuming JavaScript in Blazor*
- *Chapter 7, Managing Application State*
- *Chapter 8, Consuming Web APIs from Blazor WebAssembly*
- *Chapter 9, Authenticating and Authorizing Users in Blazor*
- *Chapter 10, Handling Errors in Blazor WebAssembly*

4

Navigation and Routing

Every software consists of a set of pages, each of which is responsible for a certain task in the UI and the functionality of the app. Within this chapter, we are going to understand the routers in the Blazor WebAssembly app, including what they are, how they work, and how to define them so that users can navigate between your app pages. We will also learn how to send data between pages via the URL and query parameters. After that, we will implement a custom UI for when the user navigates to a non-existent URL. Furthermore, we will dive a bit deeper and learn about some advanced scenarios, such as reacting to URL changes, taking actions such as cancelling ongoing tasks, and highlighting parts of the UI based on changes.

In previous chapters, we learned about components and developed individual pieces; throughout this chapter, we will put these components together within pages and implement routing so we can navigate and send data between them.

In this chapter, we will cover the following topics:

- Understanding routers and pages
- Navigation and parameters
- Handling UI changes for navigation
- Reacting to navigation changes

Technical requirements

The code used throughout this chapter is available in the book's GitHub repository here:

```
https://github.com/PacktPublishing/Mastering-Blazor-WebAssembly/
tree/main/Chapter_04
```

Understanding routing and pages

With any app you use, whether it's a native mobile or Windows app or a website, you will see it is made up of a set of pages users can move between. In the scope of the web specifically, we have traditional websites and web apps that work on the server, and we have apps that run fully in the browser, such as Blazor WebAssembly apps.

With the traditional web approach, there is no actual navigation happening in the browser because what happens is that the browser sends a request to the server for a specific URL, the server comes back with the HTML, and then the browser just renders it. On the other hand, with single-page applications, the whole app is running totally in the client, and we have a single HTML page that holds the entire UI of our app. Frameworks such as Blazor, Angular, or ReactJS, for instance, are responsible for replacing part of the UI according to the redirected link with the corresponding component, and that is what we call routing.

Routing is the process of matching the body component of the app with the requested URL. As seen in the *Building layouts in Blazor* section in *Chapter 3, Developing Advanced Components in Blazor*, within the layout component, we render the @Body property and that's where the page content of a specific URL will be shown. Now, we understand components and routing, but what are pages in Blazor?

Basically, a page in Blazor is a normal Razor component but we declare a router for it at the very top using the @page razor directive. This router is the URL used to access the component as follows:

```
@page "/counter"
...
```

The /counter value defined in the previous code snippet is used to declare that the Counter component can be accessed using the URL (app_base_url/counter).

Now, let's see who is responsible for this routing experience in Blazor.

The Router component in Blazor

If you navigate to the App.razor component, which is the root of every Blazor app, you will see that the first component called Router wraps all other components and it has a parameter called AppAssembly, which we will talk about in a bit:

```
<Router AppAssembly="@typeof(App).Assembly">
    ...
</Router>
```

The Router component is responsible for reading the routes and matching them with the components when the user navigates to that route or link.

The AppAssembly parameter, which is set to the Assembly of the project, tells the Router component where to look for components that have routes. By default, the Router component will look into the project assembly only, but if you have other pages existing in another Razor Class library project, `Router` won't be able to recognize these routes. To make `Router` look for pages in many assemblies, we need to use other parameters called `AdditionalAssemblies` and pass a collection of assemblies so that `Router` will look for and register all the available routes.

The following code snippet shows an example of adding the Razor Class Library assembly to `Router` to crawl its pages:

```
@using BooksStore.Blazor.Components
<Router AppAssembly="@typeof(App).Assembly"
AdditionalAssemblies="new[] { typeof(Component1).Assembly }">
. . .
```

`Component1` represents a component existing in a Razor Class Library project. We just use it to bring the full assembly that contains a certain type. In our `BooksStore` project, all our pages will be defined in the `BooksStore` Blazor WebAssembly; the Razor Class Library project will not contain pages, just shared components.

Creating your first page

After learning about `Router` and how navigation works, let's get started by creating our first page and discover the `NavigationManager` class, which helps us write code to navigate, as well as using the `PageTitle` component introduced in Blazor with .NET 6, which allows us to change the title of the page in the browser tab:

1. Create a new component in the `Pages/UserPages` folder and name it `About.razor`. This component will be a simple page that represents some info about the project.

2. At the very top of the `About.razor` file, we need to declare a route for this component to be accessible through `{App_Url}/about` using the `@page` directive.

3. In addition to that, when the user navigates to the `About` page, we need to change the title in the browser tab, and we can achieve that using the `PageTitle` component as follows:

    ```
    @page "/about"
    <PageTitle>About</PageTitle>
    . . .
    ```

4. Finally, add some UI content to fill out some information about the app:

    ```
    . . .
    <h2>Welcome to BooksStore</h2>
    <p>BooksStore allows you to disover best titles in tech and
    more</p>
    ```

If we run the app and navigate to `/about`, we should see the following page:

Figure 4.1 – The About page

> **Note**
>
> Each page can have multiple routes, for example, the **Index** page can have `/`, `/index`, and `/home`. Therefore, `Router` will register three URLs for the same component.

The examples you have seen in this section show you just the basics of navigation and routing in a simple setup. In real-world apps, some pages are connected, and we need to send data between those pages using parameters.

Navigation and parameters

In `BooksStore`, for example, the user can see a list of books on the **Index** page, but the user should also be able to click on a book and navigate to another page where all the book details are visible. When we navigate to the details page, we need to pass the book ID to it so that the page knows the book for which the details need to be retrieved.

Luckily, Blazor provides us with powerful and flexible routing and navigation mechanisms that allow us to navigate easily and send data in the URL, either by using the route of the page itself or using query parameters. We are going to explain both in detail.

Passing parameters using the route

Parameters can be sent within the route of the page. The `Router` component is responsible for detecting those parameters and filling in the values from the URL for the parameters in the targeted component based on the parameter name.

In the next example, we are going to create a `BookDetails` component and set a route for it, just like what we did for the `About` component; however, this route should include the book ID. The `BookDetails` component will have a `string` parameter called `BookId`, and `Router` will populate the book ID from the route.

To define a parameter within the route, we use a curly bracket with the name of the parameter inside as follows:

```
@page "/book/{bookId}"
```

bookId in the route should have a corresponding parameter in the component with the same name but Router is case-insensitive, so its parameter can be called BookId. For example, if the user navigates to https://localhost:44386/book/mastering-blazor-webassembly-2022, Router will supply mastering-blazor-webassembly-2022 to the BookId parameter after navigating to the BookDetails component.

Now, let's see this in practice:

1. Create a new component within the Pages/UserPages folder and name it BookDetails.razor.

2. Declare a Router component for this page that accepts a book ID as a parameter at the top of the component:

```
@page "/book/{bookId}"
```

3. Add a code section to the Razor file and add a parameter inside it called BookId:

```
. . .
@code
{
    [Parameter]
    public string? BookId { get; set; }
```

This is all that we need to get the book ID from the URL.

4. Now, we can start writing the content and the logic of the component. We need to write the method that retrieves a book by its ID. Let's move to the IBooksService interface we created in *Chapter 1, Understanding the Anatomy of a Blazor WebAssembly Project*, and create a new method called GetBookIdAsync that takes the book ID as a string parameter and returns a Task of type Book:

```
public interface IBooksService
{
. . .
    Task<Book?> GetBookByIdAsync(string? id);
}
```

5. Now, in the LocalBooksService class in the Services folder, let's write an implementation for the previous method in the interface that retrieves a book from the local list we have in the service:

```
...
public Task<Book?> GetBookByIdAsync(string? id)
{
    var book = _allBooks.SingleOrDefault(b => b.Id ==
                                            id);
    return Task.FromResult(book);
}
```

6. Now that we added the method to the service, let's inject it into the BookDetails component and call the function to retrieve the book by the ID passed in the URL. We can achieve this by overriding the OnInitializedAsync method that is triggered when the component is initialized. The component code should look like the following:

```
@page "/book/{bookId}"
@inject IBooksService BooksService

@code
{
    [Parameter]
    public string? BookId { get; set; }
    private Book? _book = null;
    protected async override Task OnInitializedAsync()
    {
        _book =
            await BooksService.GetBookByIdAsync(BookId);
    }
}
```

7. Next, before we design the UI, let's override the OnParametersSet method that gets called after the parameter values of the components are set. This override will help us to print the BookId parameter's value in the console window of the browser:

```
...
protected override void OnParametersSet()
{
    Console.WriteLine($"The book id is {BookId}");
}
...
```

8. Last but not least, we write the component to render the book details:

```
@inject IBooksService BooksService
...
<PageTitle>Book Details | @_book?.Title</PageTitle>
<h2>@_book?.Title</h2>
<p>@_book?.Description</p>
<ul>
    <li>Page: $@_book?.Price</li>
    <li>Author: @_book?.AuthorName</li>
</ul>

@code
{
...
```

The preceding code denotes the basic design for rendering the details of the book. Now, we can move to the next part of the process, which is the programmatical navigation to this page.

After we develop the new BookDetails page, the next phase is to navigate the user to it from the Index component when the user clicks on any BookCard component. Here is where the NavigationManager service comes into play. It's a built-in service in Blazor that provides you with the NavigateTo method, in addition to a set of helper methods for retrieving and manipulating the URL and the parameters inside it.

In the following steps, we will be using NavigationManager in the BookCard component we created in *Chapter 2, Components in Blazor*, to navigate to the BookDetails code, so let's get started:

1. Within the Shared folder, open the BookCard.razor component and inject the NavigationManager service at the top as follows:

```
@inject NavigationManager Navigation
...
```

2. In the @code section of the component, create a new method and call it NavigateToBookDetails, and inside it, we need to navigate to /book/Book_Id:

```
...
@code
{
    private void NavigateToBookDetails()
    {
        var url = $"/book/{Book.Id}";
        Navigation.NavigateTo(url);
    }
}
```

3. Finally, we just need to call the previous method when the user clicks on any point on the card.

4. In the parent `div` of the card, set the `NavigateToBookDetails` method to the `@onclick` event as follows, and you are ready to go:

```
@inject NavigationManager Navigation

<div style="@BackgroundStyle"
  @attributes="UserAttributes"
      class="card"
      @onclick="NavigateToBookDetails">
...
```

We are ready now to test what we have accomplished, so from the **Index** page, if we click on any `BookCard` listed, the app should redirect you to the **BookDetails** page, and the ID of the book should be printed in the console window of the browser, as shown in the following screenshot:

Figure 4.2 – The BookDetails page

Route parameter special conditions

What we have achieved is the ideal case; here is a list of special conditions and possibilities available for the router parameters:

- **Multiple router parameters**: Blazor allows you to set multiple parameters in a single route and the following URL shows how to achieve it:

```
@page "/profile/{id}/{onlyPosts}"
```

The previous route shows that the component takes two parameters: one called `Id` and the other one called `OnlyPosts` as an example of multiple parameters.

- **Parameter constraints**: Since the URL is a pure string, a value such as `true` can be set to a `string` parameter value or a `bool` value. Blazor provides a great feature that allows you to define explicitly what type of parameter is in the route directly. If we take the same route from the previous example of the profile, and we need to specify explicitly that `Id` is `int` and `OnlyPosts` is the `bool` value from the URL, we can declare the same route as follows:

  ```
  @page "/profile/{id:int}/{onlyPosts:bool}"
  ```

  Of course, the component parameters must have the same type declared in the route.

  Blazor supports the following C# types for route parameters: `bool`, `datetime`, `double`, `float`, `guid`, `int`, `long`, and, of course, `string`, which is the default.

- **Optional parameters**: By default, if a route parameter value is not supplied in the URL, the user will be redirected to the not found UI. Some parameters can be optional, a very common scenario in real-world apps.

- Back in our profile example, the `Id` value is required, but the `OnlyPosts` `bool` parameter can be optional and has default values of `true` or `false` set in the component parameter declaration.

  To mark a route parameter as optional, we use `?` with the name of the parameter in the route, as follows:

  ```
  @page "/profile/{id:int}/{onlyPosts:bool?}"
  ```

  Or if the type of the parameter is not defined, it could be like this:

  ```
  @page "/profile/{id:int}/{onlyPosts?}"
  ```

Now that we have discovered the route parameters, let's move to the other type of URL parameter: query string parameters.

Query string parameters

Query strings are basically a set of key-value pairs defined in a URL that assigns values to a set of parameters. The following URL shows how query strings are defined:

```
https://example.com/search?query=blazor&fromYear=2022
```

After `?` in the previous link, you can see `query=blazor`, which is a parameter called `query`, and its value is set to `blazor`. Similarly, the value of `fromYear` is `2022`. Parameters can be combined using the `&` operator.

In .NET 6.0, Blazor started support for query string parameters, and we can add as many parameters as we want to our component using the `[SupplyParameterFromQuery]` C# attribute; no changes are needed in the route of the page. Also, the name of the component parameter should not match the name in the query string parameter in the URL. The [SupplyParameterFromQuery] attribute has a Name property, which you can use to define the name of the parameter in the URL.

Let's do an experiment to see how this works.

Back in the famous Counter component, we will add a parameter that sets the default value of the _currentCount variable in the components, and the value of this parameter will use a query string from the URL. The _currentCount default value is 0, but with the parameter we will add, the initial value can be set optionally from the URL like this:

```
app_base_url/counter?currentCount=10
```

To achieve what have explained, let's go through a few simple steps:

1. Open the Counter.razor component within the Pages folder.

2. In the @code section, add an int? nullable parameter and decorate it with the [SupplyParameterFromQuery(Name = "currentCount")] attribute, in addition to the [Parameter] attribute; you can give the parameter any name you want:

    ```
    . . .
    @code {
        [Parameter]
        [SupplyParameterFromQuery(Name = "currentCount")]
        public int? CounterInitialValue { get; set; }
    . . .
    ```

    The name of the query parameter must be currentCount, equal to the value of the Name property of the [SupplyParameterFromQuery] attribute.

3. Override the OnParameterSet method and set the value of the private currentCount int variable in the component to the value of CounterInitialValue if supplied:

    ```
    . . .
        protected override void OnParametersSet()
        {
            if (CounterInitialValue != null)
                currentCount = CounterInitialValue.Value;
        }
    . . .
    ```

Now, let's run the app and navigate to the /counter page. By default, you will notice the current count in the UI starts from **0**, but if we navigate to /counter?currentCount=10, you will notice the value of the current count will start from **10** as shown in the following image:

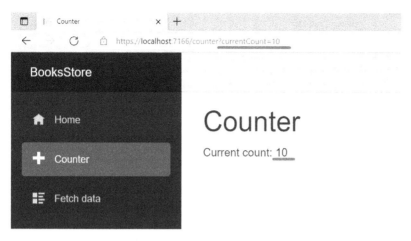

Figure 4.3 – Counter page with query string parameters

The real-world applications of query string parameters are endless. They are basically everywhere in web apps. An example is a page that renders a list of data (books, for example), for which you can use query parameters to define the query that the user is searching for, the number of items on each page, and the number of the page.

We are going to use them in many places throughout this book to send data between pages.

When you have a page that has one or more query parameters, building the URL to navigate to that page may seem a bit tricky because you could miss & between the parameters or =, as it's a pure string and that could break the parameters and their values. The NavigationManager service provides us with two helper functions, GetUriWithQueryParameter and GetUriWithQueryParameters, that can help you build a URL with a single or multiple query parameter.

The following code retrieves the URL of the app with a query parameter called search and has the Blazor value assigned to it:

```
var url = Navigation.GetUriWithQueryParameter("search", "Blazor");
```

The output of the url variable will be as follows:

```
https://app_base-url?search=Blazor
```

The preceding example retrieves a URL with a single query parameter; the following one will use the GetUriWithQueryParameters method to retrieve a URL with many query string parameters as follows:

```
var url = Navigation.GetUriWithQueryParameters("search", new
Dictionary<string, object?>
        {
            { "search", "Blazor" },
```

```
                { "pageIndex", 1 },
                { "pageSize", 10 },
                { "activeOnly", true },
            });
```

The preceding code will set the following value for the `url` variable:

```
https://Base_URL_Of_
App?search=Blazor&pageIndex=1&pageSize=10&activeOnly=true
```

Query string versus route parameters

Both string and route parameter types allow data transition via the URL, and both make you achieve what are you aiming for. Route parameters are tied to the URL and have predefined orders and shapes. However, the parameter is not human-readable, as it is declared internally as part of the URL and not using the key-value approach. Thus, it is recommended to use the parameter only when it is an actual part of the URL and not supplementary. Like in the example of the `BookDetails` page, the ID of the book is required, and it's a mandatory part of the URL. On the other hand, query string parameters are very flexible and popular; you can use them whenever you want to supply your component with some initial values for its parameters.

By now, you should be able to make your app pages linked with each other. In the next section, we will handle the UI for the requested URLs in your app that are not found.

Handling a NotFound UI

The `Router` component in Blazor has two important `RenderFragment` parameters. The first one is called `Found` and the other one is called `NotFound`. You can notice their UI piece values in the `App.razor` component:

```
<Router AppAssembly="@typeof(App).Assembly"
AdditionalAssemblies="new[] { typeof(Component1).Assembly }">
    <Found Context="routeData">
...
    </Found>
    <NotFound>
        <PageTitle>Not found</PageTitle>
        <LayoutView Layout="@typeof(MainLayout)">
            <p role="alert">Sorry, there's nothing at this
                address.</p>
        </LayoutView>
    </NotFound>
</Router>
```

The NotFound UI chunk will be rendered when the user navigates to a link that is not registered in the router. So, manipulating the content within the NotFound parameter tags will allow you to set your custom content for the not found UI.

In *Chapter 3*, *Developing Advanced Components in Blazor*, we have changed the NotFound section to try out the LayoutView component. Now, we are just going to add a little div with text and description in the middle that shows the request page cannot be found, and change the layout of the NotFound portion to UserLayout instead of MainLayout as follows:

```
. . .
<NotFound>
        <PageTitle>Not found</PageTitle>
        <LayoutView Layout="@typeof(UserLayout)">
          <div style="text-align:center;margin-top:20px">
            <h3><span style="color:red">
              <strong>Oops!</strong></span> Nothing can be
                found at this address</h3>
            <p>It looks like the page your requested in the
              URL cannot be found</p>
          </div>
        </LayoutView>
    </NotFound>
. . .
```

Next, try to run the app and navigate to an unexciting link – for example, /nothing – and you should see a UI like the one in the following screenshot:

Figure 4.4 – Not found improved UI

If the design of your not found UI is complicated or you want to share it with other projects, you can either create it in a separate Razor component in the Shared folder and render it inside the NotFound section, or create it in the Razor Class Library project so that you can share it with other projects too. This approach will keep your App.razor file short and clean.

Step by step, you are learning new things to help you complete the story of your app, get it all together, and make it look shiny for your users. In the next section, we are going to see how we can react to navigation changes in the app and why we need to do so.

Reacting to navigation changes

In the logic of your apps, sometimes, you need to react to the navigation of the user by taking them to a new URL for many reasons, such as needing to highlight a certain component in the UI if the user navigates to a specific URL, or needing to execute a piece of code when the user navigates from the current page. In this section, we will learn about the NavLink component and LocationChanged within the NavigationManager service.

The NavLink Component

Blazor contains a built-in component called NavLink. This component is basically just the normal hyperlink tag (<a>) in HTML that allows the user to redirect to a specific link using the href attribute. What makes NavLink different from using the normal HTML hyperlink tag is that NavLink reacts to the URL changes, and if the current URL matches the link defined in its href attribute, it sets a CSS class called active and removes it when the URL doesn't match. That makes the NavLink component perfect to use in building nav menus because when the user redirects to a specific URL, that link gets highlighted by adding some CSS styles for it.

In the next example, we are going to modify the NavBar component that is used in UserLayout to use the NavLink component instead of the normal HTML <a> tags.

NavLink will be able to toggle the active class when the app is within the page that this link redirects to and remove it when the user navigates away. We can highlight the link by just adding a CSS class to apply some styles to the active class:

1. Open NavBar.razor in the Shared folder.

2. Replace all the <a> elements within the ul HTML element with NavLink, and remove the active class from the <a> tag that redirects to the Home page as follows:

```
...
<ul class="navbar-nav">
            <li class="nav-item">
                <NavLink class="nav-link"
                    href="/">Home</NavLink>
            </li>
            <li class="nav-item">
                <NavLink class="nav-link"
                    href="/Counter">Counter
                </NavLink>
            </li>
```

```
                <li class="nav-item">
                    <NavLink class="nav-link"
                        href="/FetchData">Fetch Data
                    </NavLink>
                </li>
            </ul>
    ...
```

Now, if you run the app, you will notice by default the **Home** link is highlighted, and if you click on the Fetch Data link, then it will be highlighted too. That's basically the result of toggling the active CSS class applied and the default style you notice is coming from the styling the **Bootstrap** referenced in the default Blazor template:

Figure 4.5 – The hyperlink is highlighted due to the URL change detector

The LocationChanged event in NavigationManager

The NavigationManager service provides a C# event called LocationChanged. This event is fired every time the URL is changed either explicitly from the browser or using the NavigateTo function from within the app itself. Also, NavigationManager passes the required arguments to the event handlers subscribed to the LocationChanged event, which are the senders of the type object. This object represents the object that fired the event and the second argument of the LocationChangedEventArgs type, which holds two properties about the navigation changes that have occurred:

- **Location**: This string value represents the new URL

- **IsNavigationIntercepted**: This bool value indicates whether Blazor detected the URL change from the browser or by using the NavigateTo method

The following example, if you follow through with it, will show you how you can subscribe and unsubscribe to the LocationChanged event:

1. Open the Index.razor component in the Pages/UserPages folder.

2. Inject the NavigationManager service into the component, as shown in the following code snippet:

    ```
    ...
    @inject NavigationManager Navigation
    ...
    ```

3. Make the component implement the IDisposable interface using the @implement Razor directive because we need Blazor to call the Dispose method to unsubscribe from the LocationChanged event:

    ```
    ...
    @inject NavigationManager Navigation
    @implements IDisposable
    ...
    ```

4. Write a function and call it Navigation_LocationChanged. This function will take two parameters. The first is of object type and called sender, and the second is of LocationChangedEventArgs type and called e. The LocationChangedEventArgs object contains the Location property and another property of the bool type called IsNavigationIntercepted.

5. For our example, all Navigation_LocationChanged will do is print the new URL in the Console window of the browser, so your method will look like the following snippet:

    ```
    ...
    private void Navigation_LocationChanged(object? sender,
    Microsoft.AspNetCore.Components.Routing.LocationChangedEventArgs
    e)
        {
            Console.WriteLine($"The location changed to
                            {e.Location}");
        }
    ...
    ```

6. Now, we have everything ready and we just need to subscribe to the method we created for the LocationChanged event so that it gets fired every time the user navigates in your app. So, add the following line in the OnInitializedAsync method so that we can subscribe to it when the Index component is initialized:

    ```
    ...
    protected async override Task OnInitializedAsync()
        {
    ```

```
Navigation.LocationChanged +=
    Navigation_LocationChanged;
```

`. . .`

7. The last step is to write the `Dispose` function provided in the `IDisposal` interface, which will be called automatically by Blazor when the `Index` components is disposed. So, we will take advantage of that to unsubscribe from the `LocationChanged` event because if we don't unsubscribe, we may cause a memory leak due to too many subscriptions:

```
. . .
public void Dispose()
    {
        Navigation.LocationChanged -=
            Navigation_LocationChanged;
    }
. . .
```

Now, let's run the app and monitor the `Console` window. You will notice that every time you leave the **Index** page by clicking on a book card or going to the **Fetch Data** page, a message is written, as shown in the following screenshot:

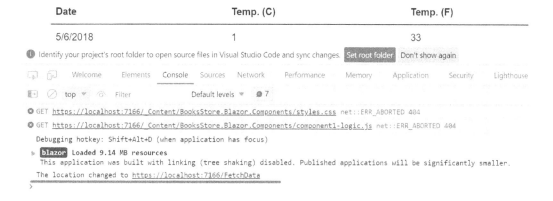

Figure 4.6 – Reaction to the LocationChanged event being fired

The NavigationLock component

The **NavigationLock** component intercepts navigation events more effectively by giving us the control to lock any navigation operation until a certain condition is met or requires a certain action before the navigation process happens.

The lifetime of the `NavigationLock` component is scoped to the lifetime of the component it's rendered inside.

`NavigationLock` accepts two parameters:

- **ConfirmExternalNavigation**: The default value is `false`, and it basically instructs the browser to prompt the user to confirm the navigation process to an external URL typed in the address bar.

  The user will be prompted only if the state of the page has changed. Let's see the following example.

  In the `Index.razor` page, add the `NavigationLock` component, and set `ConfirmExternalNavigation` to `true` as shown:

  ```
  ...
  <NavigationLock ConfirmExternalNavigation="true" />
  ...
  ```

  Now, if you run the app and try to navigate to any site, you will be redirected normally. However, if you close the modal that shows at the beginning and then try to navigate to an external site, the navigation will be locked until you confirm the operation, as shown in the following screenshot:

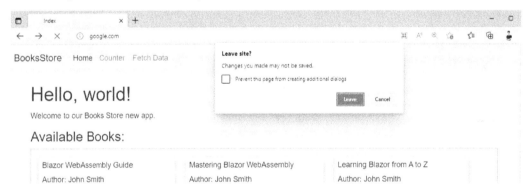

Figure 4.7 – Dialog prompts the user to confirm the navigation

- **OnBeforeInternalNavigation**: A callback that allows you to execute some logic before the internal navigation process occurs. This is useful to ask the user to save changes or wait until a certain operation is finished before leaving the page. The method to assign to the callback must take a parameter of the `LocationChangingContext` type; the object passed contains all the metadata about the navigation process.

We will be using this callback in *Chapter 5*, *Capturing User Input with Forms and Validation*, when we build a page for the admin to insert a new book.

By understanding the points mentioned, you should have more flexibility in controlling the process of your app development with Blazor and have wider knowledge to handle certain cases and problems that you may face along the way.

Summary

In this chapter, we introduced and applied all the concepts related to navigation and routing in a Blazor WebAssembly application. We started first by understanding routing and the `Router` component, and then we declared our first component that had a router. After that, we saw how to use the routing abilities in Blazor to transmit data between the pages using route and query string parameters. Then, we learned how to handle the not found UI and designed a simple and neat component for it.

Finally, we saw how we can take action and manipulate the UI when a navigation process happens in an application.

After going through this chapter, you should be able to do the following:

- Create pages in your application and declare routes for them
- Send data between pages using the route and query string parameters
- Provide a custom UI for the not found address
- Take certain actions when a navigation process occurs in the app

In the next chapter, we will start exploring a very interactive topic, which is asking the user for data input using forms, learning how to validate this data, and developing a custom input component by leveraging the `InputBase<T>` class.

Further reading

- *Routing in Blazor*: `https://learn.microsoft.com/en-us/aspnet/core/blazor/fundamentals/routing?view=aspnetcore-7.0`

5

Capturing User Input with Forms and Validation

So far, we have learned how to develop components, but all of the components we have built are for displaying data and rendering the UI of the app. Now, we will learn about existing components and develop new ones to build forms that allow the user to insert data and validate it. In almost every app you build, you will need to ask your users for some input. Whether you are building a social network platform, a point-of-sale app, a chat app, or even a simple blog, user input is always required, either to submit a new post, insert a new sales record, or submit a comment on a blog post.

In this chapter, we will cover everything you need to know to build simple and advanced forms. We will discover how Blazor deals with forms, in addition to the different input components that are built into Blazor. After building a form, we will deep-dive into the different methods of validating the data provided by the user. Finally, we will see an example of how to develop a custom input component if your needs are not fulfilled by Blazor.

By the end of this chapter, and after going through the submission form example, you should be able to utilize Blazor capabilities to develop forms with different kinds of input components. Combining what you will learn in this chapter with what you have already learned so far, such as how to build a modal popup, will provide you with the necessary comprehensive knowledge and hands-on experience to cover most of the scenarios you will face with your apps.

In this chapter, we will cover the following topics:

- Understanding forms in Blazor
- Using Blazor's built-in input components
- Validating form data before submission
- Developing a custom input component

Technical requirements

The code used throughout this chapter is available in the book's GitHub repository at `https://github.com/PacktPublishing/Mastering-Blazor-WebAssembly/tree/main/Chapter_05/ChapterContent`.

The modified and final version, which contains a real-world implementation of forms and validation, is in the same repository in the `Final` folder: `https://github.com/PacktPublishing/Mastering-Blazor-WebAssembly/tree/main/Chapter_05/Final`.

Understanding forms in Blazor

Forms are basically sets of one or more input controls, such as text, numbers, and dates. Data collected through forms is submitted at once to achieve a task such as adding a new product to the stock. Let's look at an example of a social network platform: users should be able to add new posts or edit existing posts. In addition to that, users must be able to comment on other users' posts. To achieve all of that, you need a mechanism for collecting post interaction (insertion or deletion) data through the UI. This data collection is made possible through forms.

Another example where forms are useful is a job-seeking platform where companies publish their vacancies and job seekers search and apply for them. A form is needed so that HR users at the companies can submit their job positions with information such as descriptions, availability, conditions, and salaries. On the other hand, the user needs a form to pick a job to apply for, upload their CV and covering letter, enter the expected salary, and so on.

Forms in HTML

Forms are utilities that allow you to build one or a group of input elements to collect a set of data to accomplish a certain task. Traditional web apps already have the concept of forms in the `<form>` tag element in HTML, but before we deep-dive technically, let's discuss the benefits of forms.

In normal web apps that are built with HTML, JavaScript, and CSS, the `<form>` tag combines a set of inputs. The form provides a submission mechanism through either the *Enter* key on the keyboard or a `submit` input. The submission process validates the input and, if it is valid, it sends the data from the input as a set of key-value pairs to a specific endpoint in the server.

To explain that in a bit more depth, let's take a look at the following HTML code snippet, which defines a simple login form:

```
<form action="/login" method="post">
    <input type="email" name="email" />
    <input type="password" name="password" />
    <input type="submit" value="Login" />
</form>
```

In the preceding code, we have a `<form>` tag that takes two attributes, `action` and `method`. The action attribute has the value `/login`, which is the URL of an HTTP POST endpoint on the server that accepts the values of the email and the password needed to process the login operation. `method` is the other attribute, and it takes the value `post`, which indicates the type of the HTTP request, which could be `POST` or `GET`.

Inside the `<form>` tag, we have three HTML input elements. Each has its own type, name, and value attribute. The first one has the type `email`, which means it's a text input field. It also has certain criteria; it has to be `text` input that also has to be a valid email address, so it has built-in validation. Also, on mobile phones, when the user taps on the input to populate its value, the mobile will provide a specific keyboard to input email addresses (it has the @ button on its main keyboard, for example). The `name` attribute of the `email` input defines the key in the set of the key/values that will be sent to the server. The server expects two properties, one called `email` and one called `password`, so the browser will submit the form data as a key-value pair. You can think of that just like a dictionary, where the word is the key and the definition is the value.

The second input is of the `password` type. The browser will render it as a text field, but the input is hidden by default. It also has a *show/hide* button (which we see in every app when we log in).

The third input has the type `submit`. The browser will render it as a button, and when the user clicks on it, the browser will validate all fields inside the form. If they are valid, it will submit the data to the defined endpoint; otherwise, it will show a list of the validation errors based on the input types and the values provided by the user. The `submit` input doesn't need a `name` attribute because it will be used to submit the form. The other attribute that it has is the `value` attribute. The `value` attribute for the inputs indicates the actual value; for example, for the `email` input, `value` represents the email address provided as input, so if you define it in your code, the input will be shown with the provided value inside, but for the `submit` input, `value` is used to provide the text on the button.

The previous form could be submitted using the *Enter* key after the user populates the required values. The keyboard's *Enter* key will trigger the submission of the form just like the **Submit** button on the form.

> **Note**
> To find out more about the different input elements in HTML, you can refer to the following link: `https://www.w3schools.com/html/html_form_input_types.asp`.

Now that you have some idea of the basics of how forms work on the web in ways that we are familiar with, let's move on to how Blazor has modernized the use of forms.

The EditForm component in Blazor

In modern single-page applications, either using Blazor or any JavaScript framework, we have much more power and control over the process of forms and data submission compared to older technologies. Blazor provides us with a powerful component called `EditForm`. The `EditForm` component allows us to develop a data submission form just like in our previous HTML example, but here, we have more control when it comes to collecting the data, as we can write some logic before the data submission happens and react more flexibly to the results of the form submission process.

One of the biggest advantages of developing with C# is that it is a powerful **Object-Oriented Programming (OOP)** language, which makes the development process more of a simulation of the real-world problem we are trying to solve with our app. So, how is OOP related to forms in Blazor? The answer is that the `EditForm` component accepts a property called `Model` that takes an object as a value, and this is the object that will be bound to the values of the input components and the object that will be submitted if the data is valid.

The `EditForm` component provides an object of `EditContext` type as a cascading value based on the object assigned to the form. `EditContext` is passed as a cascading parameter to the child input components, and it provides functionalities to track all the metadata of the form, such as the values that have changed or whether they are valid or not. This enables greater flexibility in managing the form's operations compared to managing forms without `EditContext`.

After this little introduction, let's go through a practical example in our `BooksStore` application to explore the basics of forms in a practical way.

Our `BooksStore` application has an admin side where a user can manage the books. One of the essential features is to allow the admin to insert or upload a new book, so let's go over that in this exercise:

1. We already have a model called `Book` inside the `Models` folder of our project that represents the properties of the book, but we will create a new one that will be used to submit the data of the book, so we will create a new class in the `Models` folder and call it `SubmitBook.cs`. Let's add the following properties:

    ```
    public class SubmitBook
    {
        public string? Title { get; set; }
        public string? Description { get; set; }
        public string? Author { get; set; }
        public string? ISBN { get; set; }
        public int PagesCount { get; set; }
        // TODO: Other properties to be added in the next
        // sections
    }
    ```

    A new object of this class will be submitted in the next steps when the form is submitted.

2. Create a new page within the `Pages` folder that will contain the new book form submission, call it `BookForm.razor`, set a route for it (`/book/form`), set the browser tab's title of the page, and create a new @code section as follows:

    ```
    @page "/book/form"
    <PageTitle>BooksStore | Add a new book</PageTitle>
    ```

```
@*Component UI goes here *@
@code
{
    // Component logic goes here
}
```

3. Now, in the `@code` section, let's define a variable of the `SubmitBook` class that we created in *step 1*. We will bind the form's model to this object:

```
@code
{
    private SubmitBook _book = new();
}
```

4. It's now time to use the `EditForm` component and add some components to receive user input. So, let's design and build the following form. We are going to use the default classes of the Bootstrap CSS framework to speed up the design process as it's already used in the default template of the Blazor app:

```
...
<div class="row">
    <div class="col-6">
        <h2>Add a new book</h2>
        <EditForm Model="_book">
            <div class="form-group mt-1">
                <label>Title</label>
                <input type="text"
                  @bind-value="@_book.Title"
                  class="form-control"
                  placeholder="Title" />
            </div>
            <div class="form-group mt-1">
                <label>Description</label>
                <input type="text"
                  @bind-value="@_book.Description"
                  class="form-control"
                  placeholder="Description" />
            </div>
            <div class="form-group mt-1">
                <label>Number of pages</label>
                <input type="number" @bind-
                  value="@_book.PagesCount"
                  class="form-control"
                  placeholder="Number of pages" />
            </div>
            <div class="form-group mt-1">
```

```
                    <label>Author</label>
                    <input type="text" @bind-
                        value="@_book.Author"
                        class="form-control"
                        placeholder="Author" />
                </div>
                <div class="form-group mt-1">
                    <label>Price</label>
                    <input type="number" @bind-
                        value="@_book.Price"
                        class="form-control"
                        placeholder="Price" />
                </div>
                <div class="form-group mt-1">
                    <input type="submit"
                        class="btn btn-success" value="Save"
                        />
                </div>
            </EditForm>
        </div>
    </div>
    . . .
```

The previous code snippet has a `div` element with `class="row"` and a nested `div` element with `class="col-6"`. `row` and `col-6` are two classes from Bootstrap that make the form 50% the width of the screen. Inside these two `div` elements, we have an `EditForm` component with a `Model` parameter set to the `_book` variable. The `EditContext` object within the form will use this object for validation (more about that in the next section) and for tracking changes in the input values.

Within the form, we have a set of `div` blocks, each with the `form-group` class. This is a Bootstrap class that groups a `form` input with its label. Each block contains a label and an HTML input tag. Depending on the type of property we want to fill, we choose a specific type; for example, for the `Title` property, we chose `text`, and for `PagesCount`, we chose `number`.

In contrast with the HTML form we created earlier, we haven't used the `name` attribute of the input because here, we are binding each property of the `_book` object to a specific value for the corresponding input using the `@bind-value` for the input element. This allows a two-way binding mechanism so that whenever we change the value in the input, the value of the property within the object gets updated and vice versa.

The last `div` block contains a `submit` input to allow us to submit the form. The difference here compared to a normal HTML form is that this one won't submit the data to the server and send the values as key-value pairs. Instead, three `EventCallback` instances are fired. One, called `OnValidSubmit`, gets fired when the form is submitted and the data is valid (for this exercise, we haven't applied validation yet), and another one, called `OnSubmit`, gets fired every time you submit the form whether the data is valid or not. The third one, called `OnInvalidSubmit`, fires when the form data is invalid.

Before moving to the next point, if you run the app and navigate to `/books/form`, you should get a UI that's similar to this:

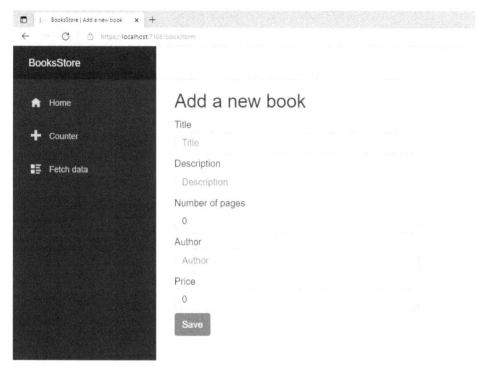

Figure 5.1 – Add a new book form UI

5. The last step in our exercise is to capture the submission of the form. This will happen when the user clicks the **Save** button or hits *Enter* on the keyboard, and to achieve that, we will handle the `OnSubmit` event callback provided by `EditForm`.

Because we don't have an API or any other data source yet, the submission of the form will just print the details entered by the user in the console window of the browser.

In the `@code` section, add the following method:

```
...
    private void HandleBookFormSubmission()
    {
        Console.WriteLine("Book has been submitted
                        successfully");
        Console.WriteLine($"Title {_book.Title}");
```

```
            Console.WriteLine($"Author {_book.Author}");
            Console.WriteLine($"Price ${_book.Price}");
    }
    ...
```

Then, let's assign the new method to the OnSubmit parameter of EditForm:

```
...
<h2>Add a new book</h2>
        <EditForm Model="_book"
            OnSubmit="HandleBookFormSubmission">
            <div class="form-group mt-1">
    ...
```

Now you are ready to give your app a run. Navigate to /book/form, try to fill out any data, and then hit *Enter* or the **Save** button. This is what you should see in your browser's console window:

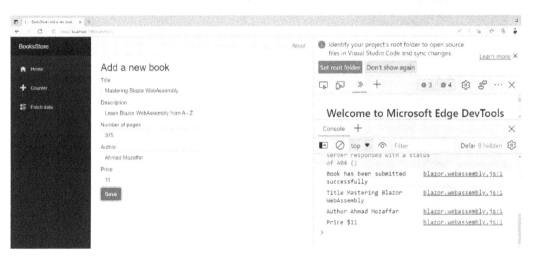

Figure 5.2 – Submit book form result in the console

Now I can congratulate you on your first form! But is that everything? Well, not quite. What if the user doesn't provide any data, or provides invalid values? We need a validation mechanism to help the user to insert valid values, but before doing that, let's discover some built-in input components that make our forms more powerful.

Note

The method that you pass to OnSubmit, OnValidSubmit, or OnInvalidSubmit can accept a parameter of type EditContext. EditForm will populate that parameter and you can use it to control the process of submission as EditContext provides you with all the metadata associated with the form, such as which field has changed and whether it's valid or not.

Discovering Blazor's built-in input components

We have a developed our first form using HTML input elements, and everything went well, but Blazor has its own set of input components that give us better control and more features. In this section, we will discover those components and upgrade our form to be more efficient.

The built-in input components automatically update the CSS class of the input based on the validation state, so we can easily apply styles for the invalid elements. We will have a closer look at that in the next section.

So, let's deep-dive into our components.

InputText

The first input component to visit is `InputText`. This gets rendered into the `<input type="text" />` HTML element. Going back to our form from the previous section, we will upgrade the input of `Title` and the input of `Author` to use the `InputText` component, as shown here:

```
...
<div class="form-group mt-1">
            <label>Title</label>
            @*<input type="text"
              @bind-value="@_book.Title"
              class="form-control"
              placeholder="Title" />*@
            <InputText @bind-Value="@_book.Title"
              class="form-control" />
        </div>
        ...
        <div class="form-group mt-1">
            <label>Author</label>
            @*<input type="text"
              bind-value="@_book.Author"
              class="form-control"
              placeholder="Author" />*@
            <InputText @bind-Value="@_book.Author"
              class="form-control"
              placeholder="Author" />
        </div>
...
```

A question arises now. Why isn't the `Description` field a text field? Actually, it is, but we will use the `InputTextArea` component for it.

InputTextArea

The InputTextArea component gets translated into a `<textarea>` HTML tag, which is a multi-line text field, and it suits the **Description** field because the user may want to insert a long description.

Back to our form, let's upgrade the input element that is bound to the Description field of the book to an InputTextArea component, as shown here:

```
...
<label>Description</label>
            @*<input type="text"
              @bind-value="@_book.Description"
              class="form-control"
              placeholder="Description" />*@
            <InputTextArea
              @bind-Value="@_book.Description"
              class="form-control" />
...
```

For our form, we still have the PagesCount and Price properties of the book, which are both numbers, so our next component is InputNumber.

InputNumber<TValue>

The InputNumber component has the same abilities as InputText, but it gets rendered to `<input type="number" />` in HTML. It's a generic component, which means we can provide the type of the value we want to deal with, such as int, double, float, decimal, and so on.

To upgrade the input used for the PagesCount property of the book, which is of type int, we can set TValue to int, as shown in the following code:

```
...
<label>Number of pages</label>
    @*<input type="number" @bind-value="@_book.PagesCount"
      class="form-control" placeholder="Number of pages"/>
    *@
    <InputNumber TValue="int"
      @bind-Value="@_book.PagesCount" class="form-control"
    />
...
```

The Price property represents a monetary value, which in software development is usually stored in a variable of the decimal type, so we can upgrade it like this:

```
...
<label>Price</label>
```

```
@*<input type="number" @bind-value="@_book.Price"
    class="form-control" placeholder="Price" />*@
<InputNumber TValue="decimal"
    @bind-Value="@_book.Price" class="form-control" />
...
```

So far, we have upgraded all our inputs for the built-in Blazor input components. If you start the application and navigate to the book form page, you won't notice a big difference. The **Description** field will have multiple lines, but if you start to enter some input, whenever you navigate from an input, you will notice a green border, which indicates that the value is valid.

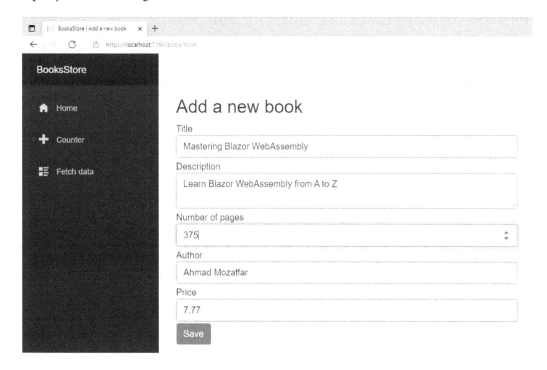

Figure 5.3 – Upgraded form with valid values in the inputs

InputCheckbox

The next component we need to know about is InputCheckbox, which is used to insert a Boolean value (true/false). It is represented as <input type="checkbox" /> in HTML and can be used like the previous inputs we have discussed (InputText, InputNumber, InputTextArea, and so on), but its value must be bound to a Boolean property, as shown in the following example, where IsPublished is a Boolean property:

```
<InputCheckbox @bind-Value="_book.IsPublished" />
```

InputDate<TValue>

Date and time are two important values too, and they have their own input component in Blazor. It's a generic component just like `InputNumber` because you can bind it to a `TimeOnly`, `DateOnly`, or `DateTime` variable type.

It can be used as shown in the following example, where `PublishingDate` is of the `DateOnly` type:

```
<InputDate TValue="DateOnly" @bind-Value="@_book.PublishingDate"
class="form-control" />
```

InputFile

`InputFile` is one of the most important input components provided by Blazor because it wraps complicated logic inside it to make it extremely easy for developers to read single or multiple files. `InputFile` gets rendered as the `<input type="file" />` HTML element, but it has an `OnChange` event, which can be used to access single or multiple files selected by the user with all their metadata using C#.

The `multiple` attribute can be used if you want `InputFile` to allow the user to choose more than one file.

To use `InputFile`, add the following component to the `component` section:

```
<InputFile OnChange="OnFileSelected" />
```

The `OnFileSelected` method must take a parameter of the `InputFileChangedEventArgs` type, and that type contains all the properties and methods needed to read and access the files and their metadata, in addition to opening a stream to read the bytes of the files.

The following code snippet contains a method that reads a single file, prints its metadata, and gets access to the file stream:

```
private void OnFileSelected(InputFileChangeEventArgs e)
{
    var file = e.File;
    Console.WriteLine($"File name {file.Name}");
    Console.WriteLine($"File size {file.Size}");
    // Access the stream of the file
    var stream = file.OpenReadStream();
}
```

To access multiple files selected by the user, you can use the following method:

```
var allFiles = e.GetMultipleFiles();
foreach (var item in allFiles)
{
```

```
    // Process each file
}
```

InputRadioGroup<TValue> and InputRadio<TValue>

Our next component is the radio button. Radio input elements allow the user to choose a single value from multiple options. `InputRadioGroup` combines a set of `InputRadio` components, and the user can select one of the `InputRadio` values, which will in turn be populated through `InputRadioGroup`.

Both components are also generic. You can bind any type of value to them because the user can select one option out of many, and this option could represent a strongly typed value in your code. This shows the power of Blazor and C#, as this technique of binding a radio input to any type of value makes the development process much easier, and you have more power while using your own C# types.

The following example shows how the user can select a book type (**Ebook**, **Paperback**, or **Both**) from a set of radio buttons and bind that value to an enum variable.

First, we need to create an enum that represents the book formats:

```
enum BookFormat
{
    Paperback,
    Ebook,
    Both
}
```

Then, we declare a variable of the `BookFormat` type and give it an initial value:

```
private BookFormat _bookFormat = BookFormat.Both;
```

Within the markup section of your component file, you need to create an `InputRadioGroup` component of the `BookFormat` type and put three `InputRadio` instances of the `BookFormat` type, each representing one of the enum values:

```
<label>Format</label>
<InputRadioGroup TValue="BookFormat"
  @bind-Value="_bookFormat">
                  <div class="row">
                      <div class="col-4">
                          <label>Paperback</label>
                          <InputRadio TValue="BookFormat"
                            Value="BookFormat.Paperback"
                          />
                      </div>
                      <div class="col-4">
                          <label>E-Book</label>
```

```
                        <InputRadio TValue="BookFormat"
                          Value="BookFormat.Ebook" />
                    </div>
                    <div class="col-4">
                        <label>Both</label>
                        <InputRadio TValue="BookFormat"
                          Value="BookFormat.Both" />
                    </div>
                </div>
            </InputRadioGroup>
```

As you may have noticed, we have bound our value or property to `InputRadioGroup`, and within it, we have three nested `InputRadio` instances, which take the `Value` parameter and represent the values that this radio button contains.

InputSelect<TValue>

`select` is a common input element in the world of software development. It represents a drop-down list that gives the user the ability to choose an element from a list, for example, choosing a country from a list of countries.

`InputSelect` gets rendered to `<select><option></option>...</select>` in HTML, where `select` represents the input and the options inside refer to each value the components contain. It's also a generic component; the selected value could be of any type.

`select` supports choosing multiple values using `multiple` attributes. In addition to that, if you want to pre-select an option programmatically, you can use the `selected` attribute of `<option>`.

The following example shows how we can add an `InputSelect` instance, which allows the user to choose the category of the book and bind the selected value into a property called `Category` of the `string` type:

```
<div class="form-group">
        <label>Category</label>
        <InputSelect TValue="string"
          @bind-Value="_book.Category"
          class="form-control">
            <option value="">Select category...</option>
            <option value="web" selected>Web</option>
            <option value="mobile">Mobile</option>
            <option value="desktop">Desktop</option>
            <option value="cloud">Cloud</option>
        </InputSelect>
    </div>
```

In the preceding code snippet, you can see that we have defined the type of the select value as `string`. Then, we bound the value to the `Category` property, which should be of the `string` type too. The actual value that the `Category` property will hold is the `value` attribute for the selected option. The text between the `option` tags is what the user sees in the UI.

If you run the app, you should get the following component:

Figure 5.4 – InputSelect component for selecting a value from multiple options

> **Note**
>
> You can see that both `InputRadioGroup` and `InputSelect` allow you to choose a value from multiple values, but the difference is quite large. Radio groups are used when selecting a single value from a small number of fixed options. On the other hand, `select` is used to choose a value or multiple values, and it's mostly used when the number of options is large. On the other hand, radio buttons are more suitable where there are few options. For example, when asking the user for their gender, either could be used, but a radio button is more suitable because there are few options, but if you are asking the user to choose a country, for example, it would make more sense to use `select`.

We have covered a wide range of input components; the next step is to make our form more efficient by validating the data inserted by the user, which ensures that the data received from the form is clean. It also helps the user to insert data that fits certain rules.

Validating form input

Validating the input means checking whether the data entered in the input components satisfies a set of predefined rules. This process has many benefits, such as ensuring the right format of data is being submitted, guiding the users to put the correct values in the correct fields, and saving resources by preventing specific logic from happening if the data is not valid. It also prevents some malicious activity that could arise from certain input.

Let's take a login page as an example. On the login page, you ask the user to enter an email address and password, then you check whether they are valid. The first thing to check is whether the email address is a valid email address and that it's not any random text value. The same applies to the password; you have rules such as the minimum number of characters and different types of characters required.

By defining these rules for the form, you ensure that you will receive the data in the correct types. It saves some resources by not allowing the user to submit the login request if the data is not valid. If the user doesn't enter any data and there is no validation, the user can click on the login button without filling out the data; the form will submit the request and it will surely fail, but by validating the input correctly, you at least ensure that you have the correct type of input.

Blazor supports validation using the C# attributes and **data annotations**, which are sets of attributes that you can define at the model level. When the user submits the form, the validation component validates the data entered by the user against the rules you have defined on your object properties.

The process of validation using data annotations occurs through a built-in component called `DataAnnotationValidator`, which can be placed within `EditForm`. By default, when the user submits the form, `DataAnnotationsValidator` accesses `EditContext` provided by the parent `EditForm` and checks the provided values against the rules defined in the model type using the data annotation attributes. After the validation process, the `EditContext` object will hold a list of the validation errors, and they can be displayed in various ways, but first, let's update our `SubmitBook` model and add some validation rules to it:

```
using System.ComponentModel.DataAnnotations;

public class SubmitBook
{
    // Title is required and value shouldn't be empty and
    // it has to be at least 3 characters long and at most
    // 80 characters long
    [Required]
    [StringLength(80, MinimumLength = 3)]
    public string Title { get; set; } = string.Empty;

    // Description is optional but it must be at most 5000
    // characters long
    [StringLength(5000)]
    public string? Description { get; set; }

    // Author is required and value shouldn't be empty and
    // it has to be at least 3 characters long and at most
    // 80 characters long
    [Required]
    [StringLength(80, MinimumLength = 3)]
```

```
public string Author { get; set; } = string.Empty;

[Range(typeof(decimal), "0", "99999")] // price decimal
// value must be between 0 and 99999
public decimal Price { get; set; }

[Range(0, 9999)] // int value must be between 0 and
                 // 9999
public int PagesCount { get; set; }

// TODO: Other properties to be added in the next
// sections
}
```

Those added attributes are all you need to build your validation logic. To make the form apply these rules before submission, we just need to add the DataAnnotationsValidator component within EditForm in the BookForm.razor file in addition to using the OnValidSubmit event callback. We should avoid using OnSubmit, as it will trigger the HandleBookFormSubmission method even if the inserted data is not valid. Thus, our validation logic should look as follows:

```
...
<EditForm Model="_book" OnValidSubmit="HandleBookFormSubmission">
        <DataAnnotationsValidator />
...
```

DataAnnotationsValidator can access the EditContext object of the form because the EditForms object supplies it to all its children as a cascading value. Thus, DataAnnotationsValidator can access the modified values entered in the form and compare them with the rules defined in the type of the object that's supplied to the Model property of EditForm.

Now, if you run the application and try to submit the form directly without inserting the correct data, the HandleBookFormSubmission function won't be triggered because the form is not valid, but the user cannot see what's wrong with it.

We have two ways to list all the validation errors for the user: either we use a component called ValidationsSummary, which lists all the errors, or we can use the ValidationMessage component, which shows the validation error for each specific property.

First, let's try to use the ValidationsSummary component by inserting it below the <DataAnnotationsValidator> component, as shown here:

```
...
<EditForm Model="_book" OnValidSubmit ="HandleBookFormSubmission">
    <DataAnnotationsValidator />
    <ValidationSummary />
...
```

It's as simple as that. Let's run the app and try to submit the form without filling in the data correctly. You should see a list of red messages that explain what's wrong with the data inside the form:

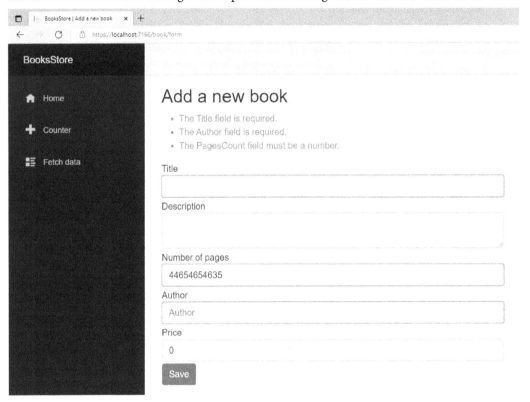

Figure 5.5 – Form submission validation result for invalid input

That's cool, right? You can also see that the outline of the field automatically turns red if the input is not valid. That's because the built-in input components attach a CSS class called `invalid` to the input, which styles the invalid components, and the default red border comes from **Bootstrap**.

`ValidationSummary` does a good job listing all the validation errors, but it would be better if we had the validation message below each input field. To do that, we can use the `ValidationMessage` component, which only shows the error for a specific property using the `For` parameter, as shown here:

```
...
<InputText @bind-Value="@_book.Title" class="form-control" />
<ValidationMessage For="() => _book.Title"  />
...
```

Now we test the app again. In addition to the list of errors at the top of the form, you will see a red message below the **Title** input field, as shown here:

Figure 5.6 – ValidationMessage for single input field output

That looks much better. Now we will comment out the `ValidationSummary` component and use `ValidationMessage` for each field to make the rest of the form look better.

The last point to mention on the topic of validation is the name of the property. Back in *Figure 5.5*, the third validation error refers to `PagesCount`, which is a programmatical name. If we want to use a more user-friendly name, in the validation annotations on the `SubmitBook` model, we can use an attribute called `DisplayName`, or we can even customize the full error message by setting the value of the `ErrorMessage` property of the attribute, as shown here in the `SubmitBook.cs` file:

```
...
[DisplayName("Number of Pages")]
[Range(0, 9999, ErrorMessage = "Number of pages must be at least 0 and
at most 9999")] // int value must be between 0
                                // and 9999
public int PagesCount { get; set; }
...
```

So far, we have got an advanced form that looks good and validates the input with a set of logical rules that make our form more efficient and provide a better user experience.

The next milestone is getting to know the APIs provided by Blazor, which allow us to develop our own customized input component in case the inputs mentioned in the previous section don't fulfill all our needs.

Developing a custom input component

Blazor contains a class called `InputBase<T>`, which is the base class for all the input components that we have seen so far. The great thing about this class is that we can use it to develop our own input component with the same specifications as the built-in ones.

`InputBase<T>` has a method called `TryParseValueFromString`, which is the main method that we need to override, and this is where we can locate the logic that processes the string of the input and translates it into something more meaningful. There is also a property called `CurrentValue`, which we can use to bind our insider input to.

Blazor already has an `InputDate` component that gets rendered to an `input` tag of the `date` type in HTML, so in the next example, we are going to create a custom component called `InputTime`, which will be translated to the `input` tag of the `time` type. We can bind a `TimeOnly` value to it. Let's get started:

1. Within the `BooksStore.Blazor.Components` Razor Class Library project, add a new component file and call it `InputTime.razor`. We have added the component to the Razor Class Library project to enable reusability.

2. Our component should inherit from `InputBase`, which exists in the `Microsoft.AspNetCore.Components.Forms` namespace, and we can set its type to `TimeOnly`. We need to implement the `TryParseValueFromString` method, as shown here:

    ```
    @using Microsoft.AspNetCore.Components.Forms;
    @inherits InputBase<TimeOnly>
    @code {
        protected override bool TryParseValueFromString(
            string value, out TimeOnly result, out string
            validationErrorMessage)
        {

        }
    }
    ```

3. Add an HTML input element of the `time` type and bind its value to the `CurrentValue` property available in `InputBase`, and also set the value of the `class` attribute to the `CssClass` property, as follows:

    ```
    . . .
    <input type="time" @bind-value="@CurrentValue" class="@
    CssClass"/>
    . . .
    ```

4. We just need to write the logic of `TryParseValueFromString` to translate the string value into a `TimeOnly` value of the output parameter result. Before we do that, we need to make sure that `value` is not empty; otherwise, we return `false` and populate the `validationErrorMessage` output parameter with a message that indicates that the process is invalid:

    ```
    protected override bool TryParseValueFromString(string? value,
    out TimeOnly result, out string validationErrorMessage)
    {
        if (string.IsNullOrWhiteSpace(value))
        {
            validationErrorMessage = "Invalid time value";
            return false;
    ```

```
            }

        result = TimeOnly.Parse(value);
        validationErrorMessage = null;
        return true;
        }
```

5. To test our new component, we can go back to SubmitBook.razor and add the new input we just created. Then, we can bind its Value to a variable of the TimeOnly type, as shown here:

```
<div class="form-group mt-1">
                <label>Time</label>
                @*<input type="number"
                  @bind-value="@_book.Price"
                  class="form-control"
                  placeholder="Price" />*@
                <BooksStore.Blazor.Components
                  .InputTime
                  @bind-Value="@_time"
                  class="form-control" />
                <p>Selected time is @_time</p>
            </div>
    ...
    @code
    {
        ...
        Private TimeOnly _time = new TimeOnly(12, 30, 0);
        ...
    }
```

That's everything we need to do. We have developed our own input class, which can be used just like InputDate, InputText, InputSelect, and others mentioned earlier, so let's run the app and see the final result of our work:

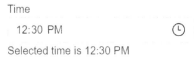

Time
12:30 PM
Selected time is 12:30 PM

Figure 5.7 – The InputTime component rendered in the UI

Because it's an input of the time type, every browser has its own time picker so the user can easily choose the time. Also, as you have seen, it's a two-way binding input. By default, the _time variable is set to 12:30:00, and that's reflected directly in the input. Also, whenever you change the input value, the variable will be updated too.

With that, you have seen another powerful capability of Blazor, which is the `InputBase` class, which we have used to develop a full input component with very few lines of code and is just like a native component.

Summary

In this chapter, we have covered some important topics in client-side software development: forms, data submission, and validation. We went over the basic definitions of forms and introduced the `EditForm` component in Blazor. Then, we reviewed the built-in input components in Blazor. After that, we covered the concept of validation and the `DataAnnotationsValidator`, `ValidationSummary`, and `ValidationMessage` components. Finally, we showed an example of how to develop our own custom input component. With these topics covered, you should have learned the following:

- Why we need forms and how they work

- How to develop a basic form using the `EditForm` component in Blazor

- How to use the built-in input components available in Blazor

- How to validate the input of the user using data annotation validators and view the result in the UI

- How to develop a custom input component for advanced cases where the built-in components aren't suitable

In the next chapter, we will continue our journey by working with JavaScript hand in hand with Blazor, seeing why and when to use it and how to achieve two-way communication between C# and JavaScript.

Further reading

- Forms and input components: `https://learn.microsoft.com/en-us/aspnet/core/blazor/forms-and-input-components?view=aspnetcore-7.0`

- `InputFile` and file uploads in Blazor WebAssembly: `https://learn.microsoft.com/en-us/aspnet/core/blazor/file-uploads?view=aspnetcore-7.0&pivots=webassembly`

6

Consuming JavaScript in Blazor

By this point, we have developed components to render data and capture the user's input and written logic and navigate between pages, all purely with C#, and that's the big promise of Blazor. But a little bit of **JavaScript** is kind of needed in some cases.

Throughout this chapter, we are going to discover those cases with some real-world examples, along with the **IJSRuntime** interface, which is a Blazor built-in service that allows communication between C# and JavaScript and vice versa.

This chapter will cover everything you need to know about JavaScript in Blazor apps:

- When and why we use JavaScript in Blazor apps
- Calling JavaScript from C# code
- Calling C# from JavaScript code
- Turning an existing JS package into a reusable Blazor component

Technical requirements

Despite the fact that this chapter focuses mainly on JavaScript, no prior knowledge of JavaScript is required to proceed, as we only use basic JavaScript code and make sure to follow it with explanations. The code used throughout this chapter is available from the book's GitHub repository:

```
https://github.com/PacktPublishing/Mastering-Blazor-WebAssembly/
tree/main/Chapter_06/ChapterContent
```

The final cleaned version of the code can be found at the following link:

```
https://github.com/PacktPublishing/Mastering-Blazor-WebAssembly/
tree/main/Chapter_06/Final
```

When and why we use JavaScript in Blazor apps

The main selling point of Blazor is that it lets us build modern web applications using C# instead of JavaScript and that's 100% true, but the presence of capabilities in the browser that have been built around JavaScript for many years makes the use of JavaScript mandatory in some cases.

Blazor already uses JavaScript behind the scenes in its engine, for example, to manipulate the **Document Object Model** (**DOM**), access some special features in the browser such as local storage, and even deal with files.

Because of the length of use of JavaScript in web development compared to Blazor, there are thousands of packages and tools that have been built using JavaScript over the years by corporates and open source contributors, including rich text editors, image croppers, and stuff like that. Having the ability to communicate with JavaScript from Blazor gives you the power to leverage those utilities directly out of the box instead of having to rewrite them in C#.

I have been using Blazor in my daily work for two and half years at the time of writing this book. The JavaScript code that I have written is minimal but nonetheless crucial, mostly used to wrap existing JavaScript libraries to be able to use them in my apps.

For example, I wrote a wrapper to integrate with the payment gateway **Stripe** (https://stripe.com). Stripe provides a collection of pre-built payment-related web elements called **StripeJS**, so I had the choice of either writing a little JS code and taking advantage of all its powerful elements, or writing my own from scratch using C# – of course, I chose to use a bit of JS.

Other times, you need just a tiny bit of JS code, only 2-3 lines, to perform some function against the DOM or trigger an event that has to be done using JavaScript. One small example here is the famous "scroll to the top from the bottom of the page" functionality that you can find on almost any lengthy web or app page. When you click this button, it scrolls the page back to the top. To achieve that behavior, a couple of lines of JS are required.

Fortunately, Blazor makes it extremely easy to enable this communication with JS in two ways, from C# to JS and from JS to C#, using the IJSRuntime interface that we can inject and get straight to work.

IJSRuntime interface in Blazor

Blazor has a built-in service that enables communication with JS called **IJSRuntime**, which is used by injecting it into any component or service when needed.

IJSRuntime has a generic C# method called **InvokeAsync** that takes as parameters the name of the JS function to be called and the set of parameters to be passed to the JS method. If the JS method returns any data, you can define the type of the data while calling the function. For example, if the JS method returns a string message, the method can be called as follows:

```
var messageFromJs = InvokeAsync<string>("SayHello", "Blazor
WebAssembly");
```

The other method to use is **InvokeVoidAsync**. This method has the exact same purpose as `InvokeAsync` but it's used to call a JS method that doesn't return data.

Blazor already takes care of serializing or deserializing the data passed to and from the JS code. For example, if the JS code requires a complex object of type book, in the C# code you just pass your book object as is, and Blazor will serialize it into a JSON object and pass it to the JS code on its own. The same applies when the JS code returns any kind of object (`string`, `number`, `Book`, etc.) – you can just define the type in C# too, as usual, without worrying how the JavaScript will become a C# object.

Before we get started writing some code that interpolates with JS, let's understand first the possible ways to reference a JS file in our applications.

Referencing JS globally

Basically, to call any JS method, the JS file that contains the method has to be referenced in your application. The traditional and preferred way is to reference the file in the `index.html` file of the app, which makes all the methods of the JS file accessible from any component.

Let's see practically how we can achieve that:

1. In the `wwwroot` folder of the application, create a new JS file named `site.js`.

2. We will write a little function that shows an alert in the browser. We will call it `showAlert(name)`, and the method will take a name as a parameter and then print a greeting for a user whenever we call this method:

   ```
   function showAlert(name) {
       alert('Hello ' + name);
   }
   ```

3. To be able to call the previous method, we need to reference the `site.js` file, so in the `index.html` page, add a `<script>` tag with the path of the file as shown here:

   ```
   <!DOCTYPE html>
   <html lang="en">
   ...
   <body>
       ...
       <script src="site.js"></script>
   </body>

   </html>
   ```

With that, the `showAlert` method can be called anytime from any component whenever we need it, as we will see in the *Calling JavaScript from C# code* section.

JS isolation in Blazor

In .NET 5.0, Blazor introduces the concept of JS and CSS isolation. Essentially, we have the ability to write some JS methods and access them on demand only through a specific scope of a JS object, and the JS resources will be loaded only when needed.

To see how this works in brief, create a new JS file in the `wwwroot` folder and give it the name `scripts.js`. Inside it, we will create a method called `getRandomBook` that returns a book object:

```
export function getRandomBook() {
    // Return a JSON object with similar properties to our
    // C# book model existing in the models folder
    return {
        title: 'Mastering Blazor WebAssembly',
        author: 'Ahmad Mozaffar',
        price: 49.99
    }
}
```

As you can see, the function is marked with `export` so it can be imported into a JS module.

To use that function, we don't need to reference the JS file in `index.html`; we can import it exactly where we want to use it.

In the next section, we will use the isolated JS technique to call the method we just created to demonstrate how to call JS methods that return data.

Calling JavaScript from C# code

Now that we have learned why we need JS and the possible ways to reference it, it's time to get some JS code executed by our Blazor app. We will cover three different scenarios:

- Calling a basic JS method
- Calling a JS method synchronously
- Calling a JS method that returns data

Calling a basic JS method

In the previous section, we created the `site.js` file that contains the `showAlert` function. In the following exercise, we will call that function from the **NavBar** component where we will add a new **Login** button on the top-right side of our app.

For now, this button will show an alert, but later we will use it for authentication purposes:

1. In the `Shared` folder, open the **NavBar** component and add a button with its own `div`, so by default, the button will show on the right-hand side of the page:

```
<nav class="navbar navbar-expand-lg navbar-light bg-light">
        . . .
        <div class="d-flex">
            <button class="btn btn-outline-primary">
                Login</button>
        </div>
    </div>
</nav>
```

2. To call the JS method, we need to inject the `IJSRuntime` interface at the top of the component file as follows:

```
@inject IJSRuntime JSRuntime
. . .
```

3. Add a code section and a method that will call the JS `showAlert` method created earlier. We will use the `InvokeVoidAsync` method as the `showAlert` JS method doesn't return any data. Otherwise, we should use `InvokeAsync` and pass to it the name of the JS method, along with another parameter, that is, the `name` parameter that the JS method expects from us:

```
. . .
@code
{
    private async Task ShowAlertAsync()
    {
        await JSRuntime.InvokeVoidAsync("showAlert",
            "Unkown user");
    }
}
```

By default, JS interop calls are all asynchronous no matter what the JS method does internally. That's why we added the `async/await` keywords to our C# method.

4. Allow the method we just created to be called when we click on the new **Login** button by using the `@onclick` event in the `<button>` tag:

```
<button class="btn btn-outline-primary" @
onclick="ShowAlertAsync">Login</button>
```

Now, if you run the app and click on the **Login** button at the top right, you should see a native alert shown in the browser with a message reading **Hello Unknown user**:

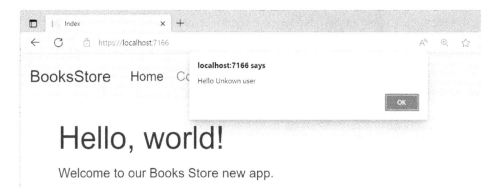

Figure 6.1 – Native browser alert

Calling a JS method synchronously

Blazor WebAssembly provides the ability to call C# methods synchronously. This allows for better control and performance in some special scenarios.

When calling a method asynchronously, the app will wait until the method finishes execution before proceeding to the following steps. However, a synchronous call will trigger the JS method and let it run in the background while proceeding to the following steps in the execution.

A great real-world use case of this is tracking events in your app. This tracking can be achieved by integrating a usage analytics platform such as *Google Analytics* into your app to track how the users are using your app. For every function the user performs in the UI, your app will send telemetry data to Google Analytics to track a specific event, and this data is sometimes sent for every mouse move. So, it doesn't make sense to execute those methods to send the telemetries asynchronously and keep the app waiting until the JS call is complete and the request has been fully sent to *Google Analytics* before proceeding with your app logic.

Lets get back to our previous exercise, where we will just alter the line that calls the JS method to make it execute the call synchronously. All we need to do is cast the IJSRuntime type into IJSInProcessRuntime and make the C# method synchronous too by removing the async/ await code as shown in the following code:

```
...
@code
{
    private void ShowAlert()
    {
        var jsInProcess = (IJSInProcessRuntime)JSRuntime;
        jsInProcess.InvokeVoid("showAlert", "Unkown user");
    }
}
```

Of course, don't forget to change the assigned function for the `@onclick` event of the button to `ShowAlert` instead of `ShowAlertAsync`.

Running the preceding code will lead to the exact same result, but the result may differ if your JS code is making some API calls as in the *Google Analytics* example.

> **Note**
>
> If your app is a pure Blazor WebAssembly app, you can safely cast **IJSRuntime** to the **IJSInProcessRuntime** instance. However, if your code will also be used as a Blazor Server app, then it will fail because Blazor Server runs on a server and requires a network connection to be able to execute the JS method, so it doesn't support synchronous calls.

Calling a JS method that returns data

So far, we have used the `InvokeVoidAsync` method to call a JS method that doesn't return anything. Next, we will go over an exercise that calls a JS method that returns an object and see how we can use it in C#.

The following exercise demonstrates a JS method that returns a JSON object representing a book. We will be using the isolated JS technique to get access to the required JS method. So, let's kick it off:

1. In the *JS isolation in Blazor* section earlier in this chapter, we created the `scripts.js` file, which contains an exportable method called `getRandomBook`.

2. Open the **Index** component that will call this method to retrieve the book object. Inject the `IJSRuntime` interface at the top:

    ```
    @inject IJSRuntime JSRuntime
    ...
    ```

3. Now we can write the code to call the method, but to do so, we need to load the JS module that holds the `getRandomBook` method. Because we are using the isolated approach, the `getRandomBook` method is not available for calling directly. First, we need to load its module and then call it from the module object in the `OnInitializedAsync` method, as shown in the following code:

    ```
    ...
    protected async override Task OnInitializedAsync()
        {
            // Call the JS import and pass the path of the
            // external js path that will be placed in the
            // wwwroot folder that will load only the JS
            // code of the requested component
            var module = await
              JSRuntime.InvokeAsync<IJSObjectReference>
    ```

```
        ("import",
         "./_content/BooksStore/Index.razor.js");
    // From the module JS object, we can call the
    // getRandomBook() method
    var randomBook = await
        module.InvokeAsync<Book>("getRandomBook");
...
```

As you may have noticed, when using `InvokeAsync` we have to define the type that we expect from the JS call. For the first call, we returned `IJSObjectReference` because this is how we import a JS module in an isolated fashion. In the second one, we used the type `Book` because the object that the `getRandomBook()` method retrieves has the same properties as our `Book` object.

Now the `randomBook` variable has the properties passed from JS and can be used normally in C#.

Calling C# from JS

When building software, a different means of communication between C# and JavaScript is required: calling C# methods from within JS code.

A good example of this is when you are trying to embed a pre-built JS library in your application, such as an image cropper or a rich text editor. In some cases, you want your app to react to changes that happen in those packages, so whenever a certain event occurs inside them, you need to fire a method in your C# code, for example, showing the dimension of the new image that the user is cropping using that JS utility.

Another example is reacting to some window changes in your app, such as when the user is resizing the window of the browser and you need to change some behaviors in your app. Maybe you have a list of emails and when the user clicks on one, it renders the email on the right-hand side. But if the user resizes the screen to make it smaller, you need the list of emails to be shown on its own page and when one email is selected, the app will redirect the user to a separate new page that will render the email.

Basically, it's not very common to implement the approach of calling C# from JS. In my 2.5 years of Blazor development I only had to do it once, while I was integrating a JS package called **GoJS**, one of the best libraries in the world to draw charts and diagrams. I used C# to JS to render the chart and set its properties using its own JS methods, but I wanted to fire some events in C# when, for example, the user right-clicked on a node in the diagram or even dragged and dropped something. For that use case, I had to use some JS to C#.

The Blazor community is huge and it's just getting bigger, but in the early days it was very common to write your own JS as it wasn't guaranteed that you would find the thing you were trying to resolve already handled by someone else. Now, it's totally different: the Blazor community is very rich and the libraries available out there cover many of our needs, especially concerning JS interop.

There are two ways of executing C# methods from JS, either calling a static C# method or calling a method of an object.

Calling a static C# method

Calling a static C# method is extremely easy. Blazor has an attribute called [JSInvokable] that we can use to decorate the required C# method so we can call it from a JS method.

If we decorate the method with [JSInvokable] alone, we can call the method using its name from JS, but [JSInvokable] takes a string as a parameter so we can give it an alias that we use when we call from it JS.

Let's go over an example and see this in action:

1. In the root directory of the project, create a demo class named JsSample.cs.

2. Create a static C# method that will just return the sum of two numbers and decorate it with the [JSInvokable] attribute as follows:

    ```
    using Microsoft.JSInterop;
    public class JsSample
    {
        [JSInvokable]
        public static int Sum(int firstNumber,
                              int secondNumber)
        {
            return firstNumber + secondNumber;
        }
    }
    ```

    To give a method an alias you can use the attribute as shown here:

    ```
    [JSInvokable("AddTwoNumbers")]
    ```

3. In the site.js file, add a function named callCsharpMethod and add the following piece of code to it:

    ```
    function callStaticCsharpMethod() {
        await DotNet.invokeMethodAsync('BooksStore', 'Sum',
                                       3, 5)
            .then(data => {
                console.log(data);
            })
            .catch(error => {
                // Handle the error here
                console.log(error);
            });
    }
    ```

DotNet is a built-in object in Blazor JS. The parameters passed to `invokeMethodAsync` respectively include `BookStore`, which is the name of the assembly that contains the static method; `Sum`, which is the name of the method or the alias name based on what you have chosen; and 3 and 5, which are the parameters required by the C# method.

After the C# method gets called successfully, you can use the `.then()` method to do after-call processing. The `data` parameter holds the output of the C# method and you can use that data within the body for any kind of logic you want. You can also call `.catch()` to handle any unexpected errors that could arise.

If you call the JS method somewhere in the app, as output it will print **3 + 5 = 8** in the browser's console window.

> **Note**
>
> The DotNet object in Blazor JS provides two methods: `invokeMethod` and `invokeMethodAsync`. The synchronous version is supported only in Blazor WebAssembly and that's why it's preferable to use the asynchronous method `inovkeMethodAsync` as it's supported in Blazor Server and WebAssembly.

Calling an instance C# method

To call an instance method, the name of the method is not enough – we need to pass the .NET object that contains that method, so it will be called for that specific object.

To achieve that, we need to pass the .NET object to the JS method, which is done by wrapping that instance with an object of type `DotNetObjectReference<ObjectType>` where `ObjectType` is the type of the instance we will pass to the JS code.

In the following exercise, we will simulate a real-world scenario where we will use the `DataListView` component we created in *Chapter 3, Developing Advanced Components in Blazor*. The `DataListView` component rendered a list of books in rows and columns, and we will add a method that will be triggered whenever the user resizes the browser window. This gives the developer a new capability based on responsiveness:

1. Open the `DataListView.razor` file in the `Shared` folder of the project. Then, inject the `IJSRuntime` interface, and implement the `IDisposable` interface at the top of the component file:

   ```
   . . .
   @inject IJSRuntime JSRuntime
   @implements IDisposable
   . . .
   ```

2. In the @code section, create a new variable of type
    DotNetObjectReference<DataListView<ItemType>>. Then, in
    OnInitalizedAsync, create its value using the Create method as follows:

    ```
    ...
        private DotNetObjectReference<DataListView
          <ItemType>>? _dotNetObjectReference;

        protected async override Task OnInitializedAsync()
        {
            _dotNetObjectReference =
              DotNetObjectReference.Create(this);
        }
    ...
    ```

    Basically, this object wraps the actual object we want to pass to the JS code, because the object
    we want to send is an instance of the DataListView<ItemType> type that will contain
    the required method we want to trigger.

3. Before we proceed with the rest of the logic, we need to call the Dispose method on
    _dotNetObjectReference when the component gets disposed to avoid any memory
    leaks that could occur. Because we are implementing the IDisposable interface, simply
    create a little method called Dispose as follows:

    ```
    ...
    public void Dispose()
        {
            _dotNetObjectReference.Dispose();
        }
    ...
    ```

4. Create a method and give it the name OnWindowResized. This method takes two int
    parameters, one for the width and the other for the height, and of course, we need to decorate
    it with the [JSInvokeable] attribute:

    ```
    ...
    [JSInvokable]
        public void OnWindowResized(int width, int height)
        {
            Console.WriteLine($"New width {width} and the
                            new height {height}");
        }
    ...
    ```

This method will be fired from the JS code whenever the user resizes the window. We can include many actions inside it, such as rearranging some elements in the UI, showing/hiding some elements, and so on, but to keep things simple, we will just print the new width and height passed from JS in the console window of the browser.

5. Back to the `site.js` file that contains the JS functions we have created so far. Create a new method in `site.js` that will subscribe to the `onresize` event and trigger the .NET function whenever the event occurs:

```
...
function triggerOnWindowResized(dotnetObjRef) {
    // Subscribe to the window.onresize event and
    // trigger a method that will trigger the .NET
    // method and pass the width and height as
    // parameters to it
    window.onresize = function () {
        dotnetObjRef.invokeMethodAsync('OnWindowResized',
window.innerWidth, window.innerHeight);
    }
}
```

The method takes a `dotnetObjRef` object that we will pass to it in the next step. Then it will subscribe to `window.onresize`, and whenever that event gets fired we will invoke the `OnWindowResized` method in the `DataListView` component by using `invokeMethodAsync`. We will pass to it the name of the method to execute and the parameters we want to pass, which in this case are the current width and height of the window.

6. For the last step, we need to call this JS method, which will call the `OnWindowResized` method. So, in the `DataListView.razor` file within the `OnInitializedAsync` method, call the `triggerOnWindowResized` JS method after creating the `dotNetObjectReference` as follows:

```
...
protected async override Task OnInitializedAsync()
    {
        _dotNetObjectReference =
          DotNetObjectReference.Create(this);
        // Call the triggerOnWindowResized JS method
        await JSRuntime.InvokeVoidAsync(
          "triggerOnWindowResized",
          _dotNetObjectReference);
    }
...
```

That's the lot. Now, because we are using this component on the index page, if we run the app and try to resize the window, we will notice every time the window size changes, some text will appear to show the new width and height in the console window, as follows:

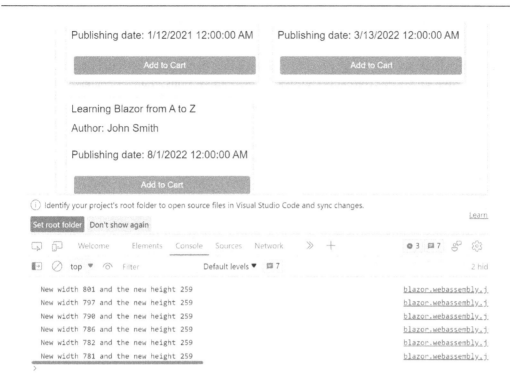

Figure 6.2 – C# method called from JS code when the window gets resized

> **Note**
> Most of the common operations that you will need are already available as NuGet packages created by the community that you can use out of the box. Still, learning how to achieve them yourself will give you more flexibility and control over the development process.

So far, we have seen many possibilities and techniques for communication between Blazor and JS. In the next section, we will work through a real-world scenario of importing an existing JS library and integrating it with our app.

Turning an existing JS package into a reusable Blazor component

JS is extremely rich in open source libraries that have been around for years and solve many scenarios. One important utility you will need in almost every app is a rich text editor. The normal text input we have in HTML only allows the user to type plain text, but with a rich text editor, the user can style their content and make it more organized and aesthetically pleasing.

In the app we are developing, we have built a form so the admin can submit a new book. In this form is a description field where the user can describe the book. In this section, we are going to leverage an open source JS library that will give us a Markdown editor that is both easy and beautiful to allow an admin writing a book description to style it and add titles, sections, and even menus.

The library is called **SimpleMDE – Markdown Editor** and is available on GitHub under the MIT license. You can find it at the following link: `https://github.com/sparksuite/simplemde-markdown-editor`. We need to write a few lines of JS to make it Blazor-ready, and because we can use it in other apps as well, we are going to place it in the `Razor Class Library` project `BooksStore.Components`. So, let's get started:

1. The library has a reference for its JS file and CSS styles that we can get from its CDN servers. We will place them in the `index.html` file of the main project as shown:

    ```
    <!DOCTYPE html>
    <html lang="en">
    <head>
        ...
        <link rel="stylesheet"
          href="https://cdn.jsdelivr.net/simplemde/latest
                /simplemde.min.css">
    </head>

    <body>
        ...
        <script
          src="https://cdn.jsdelivr.net/simplemde/latest
                /simplemde.min.js"></script>
    </body>

    </html>
    ```

2. Now, in the `BooksStore.Components` project, create a new file within the `wwwroot` folder and call it `blazor-simplemde.js`. Inside this file, write the following simple JS function, which will initialize the first `<textarea>` input in the HTML and turn it into a full Markdown editor:

    ```
    var simplemde;
    function initializeMarkdownEditor() {
        simplemde = new SimpleMDE();
    }
    ```

    I have taken the third line in the previous snippet from the library's documentation, where you can read more about this and see more advanced ways to initialize text editor input.

3. Because the text editor will be transformed into a totally new utility, we need two JS methods. One will read the content from inside the editor and the other will set the content inside it:

```
...
function getMarkdownEditorValue() {
    if (simplemde != null)
        return simplemde.value();
    return '';
}
function setMarkdownEditorValue(value) {
    if (simplemde != null)
        simplemde.value(value);
}
```

4. Create a new component in the **BooksStore.Components** project named `BlazorSimpleMde.razor`. This will be a custom component that will render a `<textarea>` HTML tag and will contain two functions to get and set the content. We use get and set methods because the values are being read from JS:

```
@using Microsoft.JSInterop;
@inject IJSRuntime JSRuntime
<textarea></textarea>
@code {
    protected override async Task
      OnAfterRenderAsync(bool firstRender)
    {
        if (firstRender)
        {
            await JSRuntime.InvokeVoidAsync(
              "initializeMarkdownEditor");
        }
    }
    public async Task<string> GetEditorValueAsync()
    {
        return await JSRuntime.InvokeAsync<string>(
          "getMarkdownEditorValue");
    }
}
```

In the preceding code, we have injected the `IJSRuntime` interface, then added a `<textarea>` tag that will be replaced by the Markdown editor. In the `@code` section, we have overridden `OnAfterRenderAsync` and called the `initializeMarkdownEditor` function from JS within `if` to check whether this is the first render. Here, we are using `OnAfterRender` to make sure that the HTML of the component is completely rendered because the HTML is not yet available in the `OnInitialized` lifecycle event.

After that, we added a method that returns a string that calls the `getMarkdownEditorValue` JS method to retrieve the string content of the editor.

5. Now, it's time to use the `BlazorSimpleMde` component in the `BookForm` component. Back in the `BooksStores` project, open the `index.html` file and add a reference to the JS file that we created. Append it to the end of the `script` section:

    ```
    . . .
    <script src="_Content/BooksStore.Blazor.Components/blazor-
    simplemde.js"></script>
    </body>
    ```

6. Open the `BookForm.razor` file from the `Pages` folder, and add a `@using` to add the namespace of the components project:

    ```
    @page "/book/form"
    @using BooksStore.Blazor.Components;
    . . .
    ```

7. Within the form, replace the `<InputTextArea>` component with the `<BlazorSimpleMde />` component:

    ```
    . . .
    <div class="form-group mt-1">
                <label>Description</label>
                @*<input type="text"
                  @bind-value="@_book.Description"
                  class="form-control"
                  placeholder="Description" />*@
                <BlazorSimpleMde @ref="_simpleMde"/>
                <ValidationMessage For="() =>
                  _book.Description"  />
            </div>
    . . .
    ```

    But because this one is not a native input component, we cannot use the binding mechanism for it. We need to fetch the value manually.

8. As you will have noticed, we created a method inside the `BlazorSampleMde` component to fetch the value from the editor, but here we are going to call a method inside a component. This is something new for us and we will achieve it using the `@ref` Razor attribut

Basically, every Blazor component in the UI is a normal C# object. Blazor provides the `@ref` attribute, which gives us access from the C# code to that instance of the component. So, in this step, we will create a C# variable of type `BlazorSimpleMde`, which we will use later in the `GetEditorValueAsync()` method:

```
...
private BlazorSimpleMde _simpleMde;
...
```

9. To populate the value of the `BlazorSimpleMde` instance we need to assign it to the `@ref` attribute of the `<BlazorSimpleMde>` component in the markup section:

```
...
    <BlazorSimpleMde @ref="_simpleMde"/>
...
```

10. The last step is to fetch the value and populate it in the `_book.Description` property. We can do this in the `HandleBookFormSubmission` method we created in *Chapter 5, Capturing User Input with Forms and Validation*, by calling the `GetEditorValueAsync` from the `_simpleMde` instance, and of course we need to mark the method as async because `GetEditorValueAsync` is an asynchronous method:

```
...
    private BlazorSimpleMde _simpleMde;
    private async void HandleBookFormSubmission()
    {
        _book.Description = await
          _simpleMde.GetEditorValueAsync();
        Console.WriteLine("Book has been submitted
                         successfully");
        Console.WriteLine($"Title {_book.Title}");
        Console.WriteLine($"Author {_book.Author}");
        Console.WriteLine($"Price ${_book.Price}");
        Console.WriteLine($"Description
                         {_book.Description}");
    }
...
```

That's everything – now we can run the app and navigate to /Book/Form and you will notice that the **Description** field has a big, advanced text editor that gives the admin user great flexibility when populating the description of the book before publishing it:

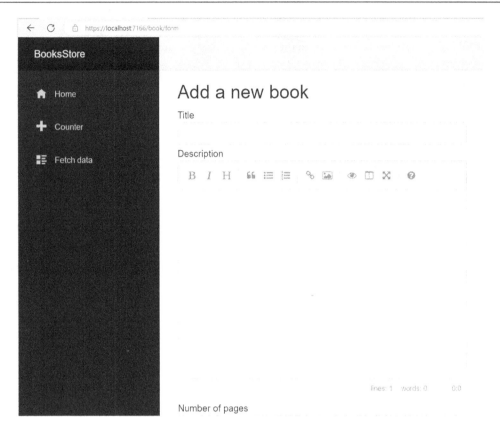

Figure 6.3 – Using the JS SimpleMarkdown Editor open source package in Blazor

As you have seen, the main language to build your apps with is C#, but whenever you need anything from JavaScript, Blazor has you covered. It puts you in a powerful position where you can leverage all your .NET skills to build rich and modern applications in addition to taking full advantage of what JS offers, all in one place.

Summary

With what we've covered in this chapter, you should be equipped with all the knowledge and experience you need to deal with anything JS related in your application, be it taking advantage of some built-in JS capabilities or using pre-built packages in your app with ease.

Throughout this chapter, we have learned why and when to use JavaScript in our Blazor WebAssembly application. We discovered the possible ways to implement interaction between the two languages in both directions. To put everything together and show you how you can leverage JavaScript to its full potential in your app, we ended up importing an open source library, wrapping it, and making it Blazor-ready so we can use it just like any other Blazor component.

By completing this chapter, you should now be able to do the following:

- Identify the purpose of JS in Blazor

- Detect when it's necessary to use some JS in your apps

- Call JS methods from your .NET code and vice versa

- Wrap a popular JS package and start using it in your apps

Further reading

- JS Interop: `https://learn.microsoft.com/en-us/aspnet/core/blazor/javascript-interoperability/?view=aspnetcore-7.0`

- Call JS from .NET: `https://learn.microsoft.com/en-us/aspnet/core/blazor/javascript-interoperability/call-javascript-from-dotnet?view=aspnetcore-7.0`

- Call .NET from JS: `https://learn.microsoft.com/en-us/aspnet/core/blazor/javascript-interoperability/call-dotnet-from-javascript?view=aspnetcore-7.0`

- JSImport/JSExport interop: `https://learn.microsoft.com/en-us/aspnet/core/blazor/javascript-interoperability/import-export-interop?view=aspnetcore-7.0`

7

Managing Application State

While a user is using an app, there is a high probability that the user will close the browser by mistake, move to another app, or even expect the app to be in the same flow after refreshing the page or opening the link in another tab.

Managing state is about keeping your app and the user experience consistent, bulletproof, and enjoyable. It's about keeping track of what the user was doing, even if something unexpected happens.

This chapter teaches you about state management, and the various ways to store and transfer this state, either by using local storage, the URL, or in-memory solutions.

While learning about state management techniques, we will apply them directly to our `BooksStore` project to preserve the state of the navigated page on the **Index** page. In addition, we will learn how to keep the state of the new book form, so that if the admin device faced an issue and the app closed suddenly, when the admin navigates to the form again, the entered data will stay there.

In this chapter, we will cover the following topics:

- What is state management?
- Persisting the state in the browser storage
- Persisting the state in the memory
- Persisting the state using the URL

Technical requirements

The code used throughout this chapter is available in the book's GitHub repository:

`https://github.com/PacktPublishing/Mastering-Blazor-WebAssembly/tree/main/Chapter_07/ChapterContent`

The finalized and cleaned version is available at the following URL:

`https://github.com/PacktPublishing/Mastering-Blazor-WebAssembly/tree/main/Chapter_07/Final`

What is state management?

When consumers use your software, they perform certain tasks that the application enables, such as adding items to a shopping cart, filling in forms, dragging events to a calendar, and more. These kinds of tasks produce data, and this data can be stored permanently (in a database on the server via an API; you can learn more about that in detail in *Chapter 8, Consuming Web APIs from Blazor WebAssembly*) or we can hold this state temporarily in a local source. In this chapter, we are focusing on the second option, which is storing the data temporarily, but before we proceed with doing that, let's understand the difference between storing on the server or locally, in addition to when to do so and why.

Basically, your software produces data, and this data can be stored in a database on the server. In the case of our bookstore, we need to store books on the server and retrieve them from there.

While developing with Blazor WebAssembly, your app needs to communicate with a Web API, which handles the communication with the data store (such as a relational database, NoSQL database, key-value storage, etc.). Developing an API that interacts with a database is out of this book's scope, but in *Chapter 8, Consuming Web APIs from Blazor WebAssembly*, you will learn how to consume an already built and published Web API to store and retrieve data about our `BooksStore` project.

Another scenario to consider is the following: if the admin is filling out a form to add a new book and then suddenly switches to another app or closes the browser after filling in that long form without saving it, does that mean they will have to fill it in again? Well, the typical answer is yes because closing the form without saving it means losing the data, and that's fine for small and short forms, but if the user opens the form again and the data is still magically there waiting for the **Save** button to be clicked, how cool would that be?!

That's one reason for keeping the state of the data within the app. Another reason is to keep the components synchronized – for example, if the user adds an item to the shopping cart, the item page that contains the **Add To Cart** button has to update the Shopping Cart component and increase the number of items in the top bar icon instantly so that the state of the application is consistent.

In *Chapter 4, Navigation and Routing*, we saw how we can use the URL and query parameters to store and send data between components; that's also another technique to keep the state of the app, and we will look at it again in this chapter but from a data perspective.

In the upcoming three sections, we will learn about and implement persisting the state of the app using three techniques: browser storage, the in-memory approach, and the URL.

Persisting the state in the browser's local storage

Most modern browsers have built-in mechanisms for storing data. This storage is **local storage**, which is used to store the data for the application across tabs. Even after closing the browser and opening it multiple times, the data stays there.

Another option is **session storage**, which stores the data for a specific session in a single browser tab. If the user reloads the tab, the data persists, but each tab has its own data for the app and the data is not shared across tabs. Also, browsers have a built-in web API to store a significant number of objects in the JSON format, called **IndexedDB**. This is an additional option for local and session storage and a very good choice for storing records of data and querying them offline during the app's lifetime.

Now, let's go over a practical example of storing the data in local storage.

In the following practice, we are going to implement a great feature in our `BooksStore` app, which is saving the new book details that the admin is filling in on the `BookForm` page.

To add a new book, we have developed a form that contains a set of properties, such as the title, description, number of pages, author, and price. Some are short fields and some are long, such as the description (which could be a long paragraph). What we will do is save what the user has filled in every 10 seconds, so if anything unexpected happens or the PC shuts down for some reason, the user will only lose up to 10 seconds' worth of work.

This information will be stored in the local storage of the browser. Blazor WebAssembly doesn't have an out-of-the-box way to communicate with the local storage, so we will be using a famous third-party library called `Blazored.LocalStorage` to achieve this.

In real-world modern applications, such a feature is a powerful part of an app, and it makes users feel safer while they are working, especially when the information is substantial, such as an application for university submissions, adding new products, writing blogs, and so on. So, always try to keep these kinds of features in mind early in the development process, as it makes your life and the user's life much easier.

Before we get started, there is a little improvement required to be able to implement the feature perfectly. In *Chapter 6*, *Consuming JavaScript in Blazor*, we developed the `BlazorSimpleMde` component, which we used as a text editor for the `description` field, and we implemented functionality to read its inner text.

Before we proceed with the `BookForm` component, we will add a property called `Content` of the `string` type to the `BlazorSimpleMde` component. We will call the `setMarkdownEditorValue` JS function that we created to populate the content of the editor initially, as right now, the initial state of the editor is always empty and there is no way to initialize it with a value.

We will do that so that if there is an unsubmitted form in the local storage of the browser, we can fetch the value and populate it in the `SimpleMDE` editor.

Now, let's see that in action:

1. Open the `BlazorSimpleMde.razor` component in the `BooksStore.Blazor.Components` project, and within the `@code` section, add a parameter named `Content` of the `string` type:

    ```
    . . .
    [Parameter]
    public string? Content { get; set; }
    . . .
    ```

2. In the `OnAfterRenderAsync` method of the `BlazorSimpleMde` component, call the `setMarkdownEditorValue` JS function if the `Content` parameter is not null or empty; after the method modification, the code should look like this:

    ```
    protected override async Task OnAfterRenderAsync(bool
    firstRender)
        {
            if (firstRender)
            {
                await JSRuntime.InvokeVoidAsync(
                    "initializeMarkdownEditor");
                // Call the setMarkdownEditorValue and
                // pass the Content as a parameter to the
                // JS function
                try
                {
                    await JSRuntime.InvokeVoidAsync(
                        "setMarkdownEditorValue", Content);
                }
                catch
                {
                    // In chapter 11, we will handle this
                    // error
                    // In real-world code avoid using
                    // catch without any logging or
                    // handling
                }

            }
        }
    ```

We wrapped the call with `try/catch` to avoid any errors that could arise from the `SimpleMDE` JS code. We will learn more about handling errors in *Chapter 10, Handling Errors in Blazor WebAssembly*.

3. Open the NuGet packages manager in the `BooksStore` project and install the `Blazored.LocalStorage` package, which is a well-known package for dealing with the local storage of the browser:

Figure 7.1 – The Blazored.LocalStorage NuGet package

You can also install it using the following `dotnet` CLI command:

```
dotnet package install Blazored.LocalStorage
```

4. After installing the package, we need to register the service in the dependency injection container by calling the `AddBlazoredLocalStorage` method. To achieve that, open the `Program.cs` file, reference the namespace of the package, and call the method:

```
...
using Blazored.LocalStorage;

...
builder.Services.AddBlazoredLocalStorage();
...
```

Now, we are ready to start using it in the `BookForm` component.

5. Open the `BookForm.razor` component in the `Pages` folder, and at the top, add the reference to the `Blazored.LocalStorage` namespace, inject the `ILocalStorage` service, and implement the `IDisposable` interface (to be used to dispose of some resources we will create later):

```
...
@using Blazored.LocalStorage
@inject ILocalStorageService LocalStorage
@implements IDisposable
```

6. In the @code section, write a method to store the current _book object, but before we call the save method, we will fetch the Description value from the BlazorSimpleMde component, as the Description property in the _book object is not filled automatically due to the JS intervention required:

```
. . .
private async Task SaveFormStateAsync()
{
    // Read the description from the SimpleMDE editor
    _book.Description =
      await _simpleMde.GetEditorValueAsync();
    // SaveItemAsync takes a key which is the 'book'
    // and the value to store, it will be serialized
    // and stored as JSON object
    await LocalStorage.SetItemAsync("book", _book);
}
. . .
```

Because the browser's local storage is key-value-based, we need to pass two parameters (the key and the value) to store; in our case, the key is book and the value is the _book object with all its properties, which will be stored as a JSON string in the local storage.

7. The other side of the feature is checking whether there is a saved Book object in the local storage, so fetch the saved Book object and populate the _book object with the stored value, and then explicitly set the content of the BlazorSimpleMDE component of the description field:

```
. . .
private async Task CheckSavedStateAsync()
{
    // ContainsKeyAsync method check Whether there is a
    // stored key in the browser's local storage
    if (await LocalStorage.ContainKeyAsync("book"))
    {
        _book = await LocalStorage.GetItemAsync<
          SubmitBook>("book");
    }
}
. . .
```

8. The last method to deal with local storage is RemoveItemAsync, and we will use this to remove a saved book object after submitting the form, as that means this saved data is no longer needed.

So, wrap the call in a method called ClearSavedStateAsync as follows:

```
. . .
private async Task ClearSavedStateAsync()
{
```

```
        await LocalStorage.RemoveItemAsync("book");
    }
    . . .
```

9. It's time to build the timer, which will trigger the SaveFormStateAsync method we created in *step 6*, so we need to create an object of the Timer type called _timer. Then, within a method called SetupTimer, we will initialize it, set its interval to 10 seconds, use its Elapsed event to trigger the SaveFormStateAsync method, and finally, call its Start method so it starts ticking:

```
    . . .
    private System.Timers.Timer _timer = new();
    private void SetupTimer()
    {
        var second = 1000;
        _timer.Interval = 10 * second;
        _timer.Elapsed += async (sender, e) =>
        {
            await SaveFormStateAsync();
        };
        _timer.Start();
    }
    . . .
```

10. To avoid any memory leaks and stay efficient, we need to dispose of the _timer object we created, and we will implement that in the Dispose method available in the IDisposable interface as shown here:

```
    . . .
    public void Dispose()
    {
        _timer.Stop();
        _timer.Dispose();
    }
    . . .
```

Now, we have all methods and logic ready enough to finalize the feature.

1. We need to override the OnInitializedAsync method so that we can call the SetupTimer and CheckSavedStateAsync methods. As a result, whenever the user redirects to the BookForm page, the timer starts ticking to save the state every 10 seconds and also checks whether there is saved data in the local storage to fetch and populate it:

```
    . . .
    protected async override Task OnInitializedAsync()
    {
```

```
        SetupTimer();
        await CheckSavedStateAsync();
    }
    ...
```

2. Finally, when the user submits the form successfully, we need to clear any stored data by calling `ClearSavedStateAsync` at the end of the `HandleBookFormSubmission` method we created in *Chapter 5, Capturing User Input with Forms and Validation*:

```
    private async void HandleBookFormSubmission()
    {
        ...
        await ClearSavedStateAsync();
    }
```

Everything is in good shape right now to be tested and to make sure this functionality is working as expected, so run the application, navigate to /book/form, then start populating the fields in the form, such as the Title and Description fields.

After 10 seconds or more, refresh the page, or close the browser and open it again.

You can have a look at the data within the local storage by opening the developer tools of the browser, navigating to the **Application** section, then expanding **Local Storage**, and clicking on the URL of your app.

On the right side, you should see the book key with its value, which represents a _book object in JSON format, as shown here:

Figure 7.2 – The book object stored in the local storage

> **Note**
> There is another library for interacting with the session storage of the browser, which is `Blazored.SessionStorage`. You can learn more about it via the following link: `https://github.com/Blazored/SessionStorage`. Also, an open source package is available for Blazor WebAssembly to store and retrieve data from a browser's `IndexedDB`. You can find the steps on how to get started with it here: `https://github.com/nwestfall/BlazorDB`.

Now we have learned how to store the app state in the storage of the browser, let's find out how we can persist the state of the app using an in-memory solution.

Persisting the state in the memory

Using in-memory objects and instances is another option for keeping the state of the app synchronized across the components and instances.

We can achieve this by creating a class that has the property to be stored, a delegate so that the consumer components can subscribe to it to execute some logic when its value changes, and an update method, which sets the value of the property and triggers the delegate.

After creating the class, we register an instance of it in the dependency injection container as a **singleton**, which will be a single instance created when the app is loaded in the browser, and the same instance is used when any component or service injects the object.

One component can call the `Update` method and pass the value, and other components can be listeners for the change by subscribing to the delegate and executing some logic in reaction to the value change.

Let's take a scenario of an app that has certain dashboards, where each dashboard visualizes data for a period that the user can define – for example, the last 24 hours, the last week, and the last month.

The app has a *period picker* on its navbar at the top of the page, and whenever the user changes the period, all the dashboard pages must react to this event and keep that state across the application's lifetime.

The following diagram shows how this example works:

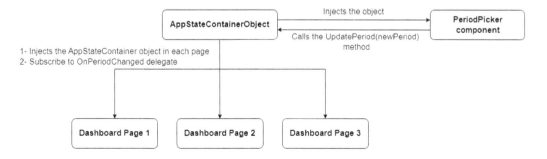

Figure 7.3 – Flow of in-memory state management

Now, let's practically demonstrate how we can achieve what we have explained. To keep things simple, we will use the **Counter** page. We will store the current counter in an in-memory object, and its value will be reflected beside the title of the page in the navbar.

When the app runs, an instance of the state object will be created, and the current counter will be initialized to 0. We will inject this object into the **Counter** page, and in the Navbar component, the **Counter** page will update the current counter value, while the Navbar component will listen to that change and update the UI accordingly. Follow these steps:

1. Let's start creating the state container object by creating a new class in the BooksStore project and calling it AppStateContainer.cs.

2. The object will hold the CurrentCounter property of the int type, a delegate of the Action<int> type called OnCounterChanged, and listener components, and can subscribe to this delegate so that whenever it gets triggered, they can receive the new value, and finally, the UpdateCounter method, which updates the CurrentCounter property and triggers the OnCounterChanged delegate:

    ```
    public class AppStateContainer
    {
        public int CurrentCounter { get; set; }

        public Action<int>? OnCounterChanged { get; set; }

        public void UpdateCounter(int newCounter)
        {
            CurrentCounter = newCounter;
            OnCounterChanged?.Invoke(newCounter);
        }
    }
    ```

3. Register this object in the dependency injection container by opening the Program.cs file and registering AppStateContainer using the AddSingleton method, which means registering a single instance of the object that lives throughout the app's lifetime:

    ```
    ...
    // The class instance will be injected in every
    // component or service that injects the
    // AppStateContainer
    builder.Services.AddSingleton<AppStateContainer>();
    await builder.Build().RunAsync();
    ```

The AppStateContainer object is ready to be used by the **Counter** page and the Navbar component. We will start with the **Counter** page first.

4. In `Counter.razor`, in the `Pages` folder, inject `AppStateContainer` at the top of the page, and set its layout to `UserLayout` so that it uses the public layout, not the default one, which is supposed to be for an admin user:

```
@page "/counter"
@inject AppStateContainer AppStateContainer
@layout UserLayout
. . .
```

5. We need to bring back the **Increment Count** button that we removed in *Chapter 4, Navigation and Routing*, so when we click it, the `IncrementCount` method is executed:

```
@* Add the Increment Count button*@
<button class="btn btn-primary" @
onclick="IncrementCount">Increment Count</button>
@code {
. . .
```

6. In the `IncrementCount` method, after the line that increases the `currentCount` variable by one, we will call the `UpdateCounter` method in `AppStateCounter`, so the full method will look like this:

```
private void IncrementCount()
{
    currentCount++;
    AppStateContainer.UpdateCounter(currentCount);
}
```

After running the app and navigating to `/counter`, the page should look like this:

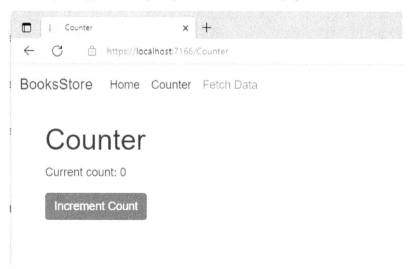

Figure 7.4 – Updated Counter page

7. The other side of the process is the `Navbar.razor` component in the `Shared` folder. We need to append the current counter value to the **Counter** link in the menu so that the **Counter** page link will be `Counter (Current_Counter_Value)`.

    Let's inject the `AppStateContainer` object, and implement the `IDisposable` interface:

    ```
    @inject AppStateContainer AppStateContainer
    @implements IDisposable
    . . .
    ```

8. In the `@code` section, create a variable of the `int` type called `currentCount`, and a method called `OnCounterChange`, which takes a parameter of the `int` type.

    The method will set the value of the local variable to the passed value of the parameter and call the `StateHasChanged` method. As this `OnCounterChanged` method will be triggered from the `AppStateContainer` object outside the `Navbar` component, Blazor doesn't update the UI accordingly because of the render tree tracking (we will talk more about this in *Chapter 12, RenderTree in Blazor*):

    ```
    . . .
    private int currentCount = 0;
    private void OnCounterChanged(int newCounter)
    {
        currentCount = newCounter;
        StateHasChanged();
    }
    . . .
    ```

9. Override the `OnInitialized` method and, within it, add the following logic:

    ```
    protected override void OnInitialized()
    {
        // Set the initial value of the currentState
        // variable to the value from the
        // AppStateContainer
        currentCount =
          AppStateContainer.CurrentCounter;
        // Subscribe to the OnCounterChanged delegate
        // and assign the OnCounterChanged method to
        // it
        AppStateContainer.OnCounterChanged += OnCounterChanged;
    }
    ```

10. Unsubscribe from the `OnCounterChanged` delegate in the `Dispose` method to avoid keeping the subscriptions in the memory, as keeping them will prevent the Garbage Collector from collecting the whole component, and that could lead to memory leak issues:

```
public void Dispose()
{
    AppStateContainer.OnCounterChanged -=
        OnCounterChanged;
}
```

11. Finally, in the markup part of the file, update the content of the `NavLink` component for the **Counter** page to add the `currentCount` value to it:

```
. . .
    <NavLink class="nav-link" href="/Counter">Counter
        @(currentCount)</NavLink>
. . .
```

Here we go! Now, we can run the app! The first thing you will notice is the **Counter (0)** tab in the navbar:

Figure 7.5 – Updated Counter page link

Navigate to the **Counter** page, and click **Increment Count** multiple times. The first thing to notice is that the **Counter** link in the navbar is updated with the latest current counter value.

The second thing is that whenever you navigate in the app, the current counter value will persist, and the latest value will stay beside the **Counter** link, as follows:

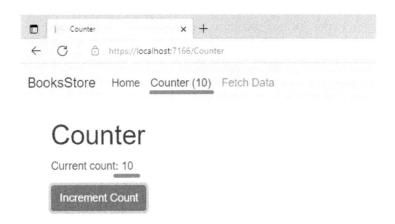

Figure 7.6 – Current counter state persists using in-memory object

After implementing this demonstration successfully, you should be able to use this solution in your projects to solve many cases, such as updating a shopping cart and keeping it stored, or even receiving and showing notifications while persisting the whole app state based on the notification count, read status, and so much more.

In this section, we have seen how we can use a local in-memory object to keep the state of the components consistent. In the next section, we will learn how to use the URL and query parameters as another technique to persist the state.

Persisting the state using the URL

The URL is used mainly for navigation between the app pages, but we have already seen in *Chapter 4, Navigation and Routing*, how we can send data in the URL to other components and pages.

The same techniques we used, such as the routing parameters or query parameters to transfer data, are also used to keep the flow and the state of the app.

A common example that you may face in every app you build is having a page where there is a list of data to show, just like the **Index** page of our `BooksStore` app, which renders a list of books.

The list can be long, so pagination is the solution; each page shows a chunk of the books, and the user can navigate between the pages.

For a good user experience and to make the app easier to use, you need to keep the state of the current page that the user is navigating. When the user refreshes the page, shares the link, or opens it in another tab or browser, the same chunk of data will be listed, and the state of the page will persist if the page number in the URL is passed as a query parameter.

This technique is used widely, and you can notice it in almost every app you open. Storing the data in a URL is a common approach and it's good to plan your components and pages for this scenario from the early stages.

In the next exercise, we will implement adding a page number query parameter to the **Index** page and see how we can save that state using the URL:

1. Open the `Index.razor` file within the `Pages/UserPages` folder, and in the `@code` section, add a parameter of the `int` type called `PageNumber`:

    ```
    . . .
    [Parameter]
    [SupplyParameterFromQuery(Name="pageNumber")]
    public int PageNumber { get; set; };
    . . .
    ```

2. Override the `OnParametersSet` method, and add a check for `PageNumber`; if it equals or is less than 0, set its value to 1:

    ```
    . . .
    protected override void OnParametersSet()
    {
        if (PageNumber <= 0)
            PageNumber = 1;
    }
    . . .
    ```

    Such a step is important to keep your app consistent and efficient and avoid any scenarios that could lead to unlikely behavior in the logic.

3. To keep things simple, we will print the number of the page in the browser's tab title, using the `<PageTitle>` component at the top:

    ```
    . . .
    <PageTitle>Index | Page @PageNumber</PageTitle>
    . . .
    ```

4. Add a method called `GoToPage`, which takes the number of the page; the method will navigate to the same URL but append the value of the query parameter as shown:

    ```
    . . .
    private void GoToPage(int pageNumber)
    {
        PageNumber = pageNumber;
        Navigation.NavigateTo(
          $"/?pageNumber={pageNumber}");
        // TODO: Logic to fetch the books by the new page
        // number
    ```

```
    }
    . . .
```

The local navigation to the same page won't achieve anything other than adding the value to the URL; if the user shares the link or refreshes it, the page number will be persisted.

5. For testing purposes, we will add three little buttons to navigate between the three pages below the `DataListView` component, as shown:

```
. . .
<div class="mt-2">
    <button class="btn btn-primary" @onclick="() =>
      GoToPage(1)">1</button>
    <button class="btn btn-primary" @onclick="() =>
      GoToPage(2)">2</button>
    <button class="btn btn-primary" @onclick="() =>
      GoToPage(3)">3</button>
</div>
. . .
```

Done! Run the app, and you should notice that the title of the tab in the browser has the page number there, in addition to three buttons. If you click on them, the page number will be added or updated in the URL, as shown here:

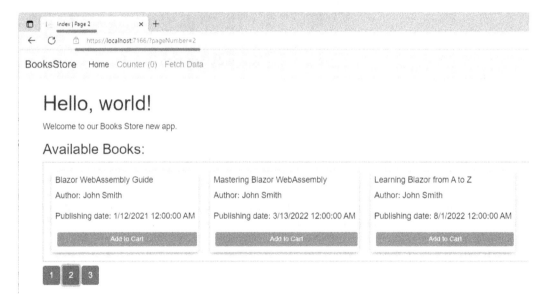

Figure 7.7 – Page number stored in the URL

In a real-world scenario, more details are needed, such as disabling the button of the current page, in addition to some design aspects to give the ultimate experience for your users.

The examples are not limited to paging. When you use any web app, notice the URL and how much data is stored there, and you can learn how to use the URL and query parameters more efficiently.

While building your own apps, try to think about state management features from the infrastructure level, as these make the app more useable and smoother to use, even though state management doesn't have any UI effects. If the `Pages` button only changes the page number internally in the component every time the page gets refreshed, the user has to navigate to the target page again, in addition to losing the possibility of sharing a link that points directly to the chunk of data needed.

> **Note**
> Modern software operates as smoothly as possible and serves the user in the easiest way possible; state management relates directly to that aspect, so take care of it very well!

Last but not least, if you are already a frontend developer with any JS frameworks such as React, you have already heard about Redux. Basically, **Redux** is a JS library that centralizes the process of state management in a single place, and it's very common in the JS world. Luckily! The great community of Blazor developers has got you covered; **Blazor.Fluxor** is a Redux-like alternative for Blazor. You can learn more about Fluxor and how to get started with it at the following link: `https://github.com/mrpmorris/Fluxor`.

Summary

In this chapter, we learned how we can keep the user experience as consistent as possible, as well as keeping it safe by persisting the state of the app.

We started by defining what state management is and then we discovered the possible ways to persist the flow of what the user was doing, from browser storage to an in-memory approach, and finally, using the URL.

Alongside the explanation, we explored three practical examples that added powerful features to our app, from storing the data the user is filling in to keep it safe for any surprise event to using the URL to store the current page number that the user has navigated to.

After going through this chapter, you should be able to do the following:

- Define the importance of state management
- Know how to use the local storage of the browser
- Know how to save the state of the app using in-memory objects and dependency injection
- Use query parameters to store the current state of the page and the flow of the user

Further reading

- Managing state in Blazor: `https://learn.microsoft.com/en-us/aspnet/core/blazor/state-management?view=aspnetcore-7.0&pivots=webassembly`

- More about the garbage collector in .NET for a better understanding of why disposing of objects and event subscriptions are important: `https://learn.microsoft.com/en-us/dotnet/standard/garbage-collection/fundamentals`

8

Consuming Web APIs from Blazor WebAssembly

It's time to feed our application and the page we have built with a real data source, let the app communicate with an online service to list the books available, and create the functionality for an admin to add new ones.

We will start this chapter by explaining why you need to communicate with an API from your Blazor app, in addition to understanding the built-in .NET class **HttpClient**, as well as other extensions that will be used to implement the communication with the API.

Next, we will practice sending GET and POST requests to our API. We will also explore some more advanced techniques to manage and manipulate HttpClient instances. We will use **IHttpClientFactory** and delegating handlers to apply global logic to HTTP requests and responses. By the end of the chapter, we will have learned how to encapsulate the API calls and take them out of the components to keep them clean and testable.

This chapter will enable you to consume APIs, either APIs you have built or third-party APIs, and teach you to implement API calls in a clear, efficient, and readable way. The following topics will be covered in this chapter:

- Understanding web API clients
- Calling a web API from Blazor WebAssembly
- Sending a GET request
- Implementing a POST web API call in Blazor WebAssembly
- Exploring IHttpClientFactory and delegating handlers
- Separating your API calls from components

Technical requirements

The Blazor WebAssembly code used throughout this chapter is available in this book's GitHub repository:

`https://github.com/PacktPublishing/Mastering-Blazor-WebAssembly/tree/main/Chapter_08/Chapter_Content`

Within this chapter, there will be a separate API project that you can clone from the GitHub repository too:

`https://github.com/PacktPublishing/Mastering-Blazor-WebAssembly/tree/main/Chapter_08/API`

The API project is required to test the calls, so after cloning the API solution, you can run the project just like running Blazor WebAssembly using either of these two options:

- Open the terminal in the API project folder and run these commands:

    ```
    dotnet build
    dotnet run
    ```

- Alternatively, run the project using the Start button in Visual Studio.

If the project runs successfully, navigate in the browser to `https://localhost:7188` and you should see the **Swagger** (API documentation) page.

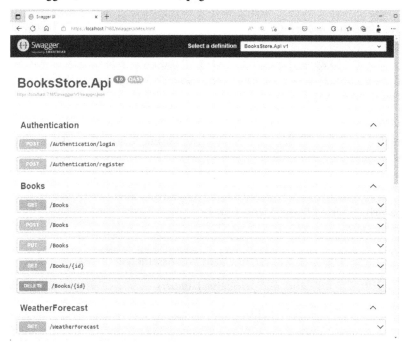

Figure 8.1 – BooksStore API documentation page

> **Note**
>
> Because this book focuses only on the client side, there will be no explanation of how the API project works internally, and all the code will focus on consuming the API instead.

Understanding web API clients

Before we start consuming the `BooksStore` API, let's talk a bit about web APIs. Basically, an API is a kind of interface for a set of functions, and this interface is accessible by code.

Let's take smartphone notifications as an example. If you are developing an app for a smartphone, be it iOS or Android, the operating system provides you with an API to deal with its notification functionality for things such as showing a new notification or defining what should happen when you click on a notification.

A web API is also an application programming interface, as the name suggests, but it's accessible over the web via HTTP. It's a kind of web application that exposes certain functionalities and is hosted on a server somewhere in the world. Your application can consume the functions that the web API exposes by sending HTTP requests, and it will send you the status of the data received in an HTTP response.

Web APIs are a crucial part of modern software development and are almost ubiquitous. When you open your smartphone to check the weather, the app is a client that renders weather data coming from a specific web API. Likewise, when you navigate the news in your browser and see many articles from different sources in one place, that's all facilitated by the web APIs of the different news agencies.

No matter the framework you are working on for a client application, whether Blazor or any JavaScript framework, and whether it's a Windows, Android, or iOS client, the API is accessible via HTTP, so you don't need to be concerned with the language used to build that API or what kind of server it is hosted on.

When do you need a web API?

Well, in most cases, you need a backend for your application, no matter what the app does. The web API wraps all the logic of the system – the business logic, communication with the database, and user management, among other stuff.

In rare cases, you can have an app that runs on the client machine in the browser or a native app that doesn't require an API (for example, a Markdown editor or a drawing app), but even in these scenarios, a web API may still be needed to manage users or store their files in the cloud.

That leads to another question – do you always need to have your own API? The answer is not always. There are some scenarios in which you may develop a frontend for a third-party API – for example, building a contacts app. In that case, you can rely on the Microsoft API (Microsoft Graph) to store or retrieve user contacts. The user can log in with a Microsoft account and get access to their contacts.

However, for our bookstore project, we need a web API that wraps all the logic of storing and retrieving books, in addition to managing users and user orders. So, as a Blazor WebAssembly app developer, should you be responsible for developing the API? That actually depends on whether you are a solo developer or an employee in an organization. If you are a solo developer, you will build your own app, as you're basically a full stack developer, so you will also need to build that API yourself. If you are working in an organization, there may be separate frontend and backend teams for a project. Thus, you may only need to work on the Blazor app.

In any of the aforementioned scenarios, while you are developing a Blazor WebAssembly app, you will consume a web API, whether it's a third-party web API, an API you developed, or one developed by your colleagues.

Throughout this chapter, you will learn how to understand and read an existing API, use utilities such as **Postman** to test it, and then leverage the `HttpClient` class in .NET to send requests to the web API you are working with.

Understanding and reading a web API

As a frontend developer, you need to be able to understand the API you are dealing with by identifying requests and their schemes, the parameters they accept, the payload you need to send, and the response that the web API returns for each request.

Each web API exposes a set of methods known as endpoints, which you call by sending an HTTP request that points to that endpoint alongside the required request metadata and body if needed. After you send the request and the web API processes it successfully, it will send you an HTTP response that also contains a status code describing the status of your request, along with an optional body if the endpoint you called returns any data.

Now, the question that arises is, how would you know all those details about the web API, especially if you did not develop it?

Every modern web API has documentation that lists the available requests with all their accepted parameters and the data it receives in the body if there is any, in addition to the response and the scheme of the data it contains.

For the `BooksStore` API, when you run the project and navigate to `https://localhost:7188`, you will be automatically redirected to `/swagger/index.html`, as follows:

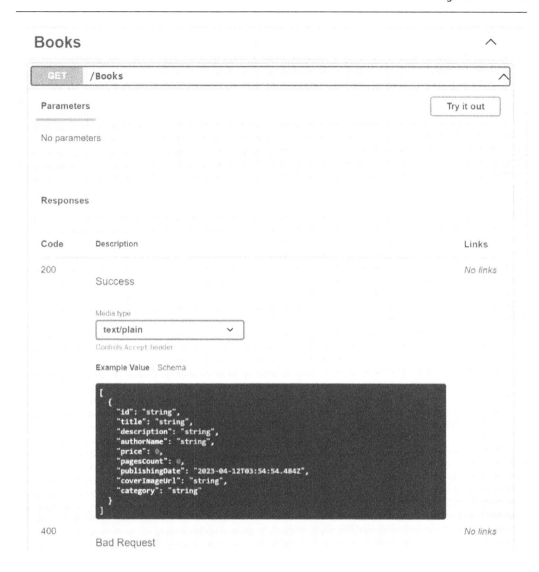

Figure 8.2 – BooksStore API Swagger documentation

The UI shown in the preceding figure is the **Swagger** page. This page is auto-generated by the Swagger framework. It reads the structure of an ASP.NET Core API and generates documentation for it. The Swagger page also allows you to test API calls directly from the page.

Most third-party web APIs, such as Microsoft Graph, have their own documentation online. You can find the full documentation for Microsoft Graph at the following link: `https://learn.microsoft.com/en-us/graph/use-the-api`. You can search online for the documentation of your targeted API and use it to kickstart your exploration and testing journey.

The `BooksStore` API exposes multiple endpoints, two for authentication (login and register) and five book actions (get all, get by ID, create, update, and delete). As you can see in the Swagger documentation, the requests are identified with the words POST, GET, PUT, and DELETE. Those are the method types of HTTP requests. The method is used to clarify the purpose of the request. The following table shows the most commonly used methods, alongside their usages:

Method	Description (Purpose)	Specification
POST	Create a new resource	The request has a body that you can send the data into
GET	Retrieve a specific resource	The GET request doesn't have a body to send data into; the data or parameters must be passed in the URL
PUT	Update or replace (replacing the full object)	The request has a body that you can send the data into
PATCH	Update or modify (update one property or more)	The request has a body
DELETE	Delete a resource	The request doesn't have a body and the targeted resource should be passed in the URL as a parameter, such as the ID of the resource

Table 8.1 – Most commonly used HTTP methods

Referring back to *Figure 8.2*, the /Books GET request is expanded, and, as you can see in the figure, the Swagger page shows you the possible responses – for example, the status code 200, which represents a successful request. It will retrieve a JSON file containing an array of books. The Swagger documentation also lists the objects and their schemes used in the endpoints at the bottom of the page. From those schemes, you can build your classes, as we will do in the next section.

Testing a web API with Postman

Before you start to write code to consume a web API, it's good practice to test the APIs with a specialized tool. Despite the fact that Swagger provides a mechanism to test the API directly from the page itself, we will use a tool called **Postman**. You can find and download it from the following link: https://www.postman.com/. Because not all APIs use Swagger, learning how to test them using Postman gives you more flexibility, especially when working with third-party web APIs.

Postman is a common and easy-to-use app that you can utilize to simulate API calls. Postman is a huge app and has a variety of features, but we will focus on sending requests and reading the response only, as that is what's needed for testing the endpoint. We will start by writing the code to send the simulated requests from our app.

You may ask why we need to use a tool such as Postman if we can already write the code directly. Well, the thing is that by using Postman, you can test the calls quickly without effort and identify exactly what URL to use, what parameters to pass, what objects to send in the body, and, finally, understand the response, and then start your coding around that.

Because our API is written in ASP.NET Core, it comes with a weather forecast endpoint by default, which is what we will test in this section:

1. Run the API and navigate to the Swagger page. You will see a /WeatherForecast GET request.

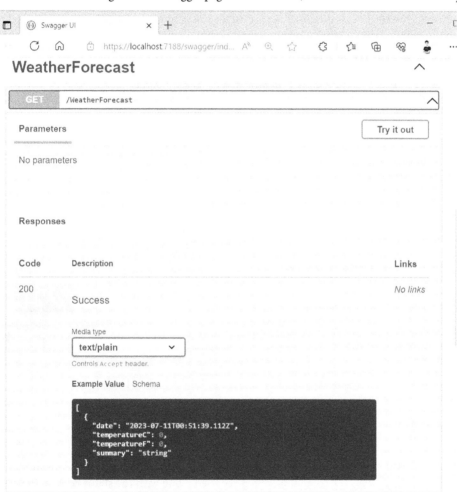

Figure 8.3 – /WeatherForecast GET request in documentation

2. After you have created an account (if you don't have one already), open Postman and click on the **New** button in the top left:

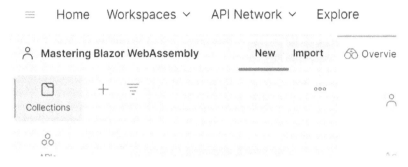

Figure 8.4 – Postman New request button

3. Choose **HTTP Request** from the dialog that appears.

4. In the new request tab, make sure to insert the base URL of the API, followed by the route of the endpoint, which is /weatherforecast. Then, choose **GET** from the dropdown on the left-hand side of the URL bar, as shown here:

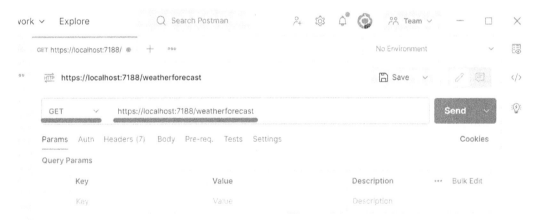

Figure 8.5 – New request in Postman

5. Click the **Send** button and Postman will send the request to the web API. You should see a response saying that the content has been received along with the status code in the bottom section of Postman's UI:

Body   Cookies   Headers (4)   Test Results   200 OK   8

Pretty   Raw   Preview   Visualize   JSON ∨   ⇥

```
 1  [
 2      {
 3          "date": "2023-02-16T04:23:33",
 4          "temperatureC": 20,
 5          "temperatureF": 67,
 6          "summary": "Hot"
 7      },
 8      {
 9          "date": "2023-02-17T04:23:33",
10          "temperatureC": -2,
11          "temperatureF": 29,
12          "summary": "Scorching"
13      },
14      {
15          "date": "2023-02-18T04:23:33",
16          "temperatureC": 5,
17          "temperatureF": 40,
18          "summary": "Balmy"
19      },
```

Figure 8.6 – /WeatherForecast endpoint successful response

Congrats! This output means you have set up the API project and Postman successfully, understood the meaning behind what we will be doing further on in this chapter, and made sure that everything works fine. Let's get started writing some code to achieve the same thing in our Blazor app.

Calling a web API from Blazor WebAssembly

Making calls to the web API from the Blazor app involves the same steps as doing so from Postman. We start by defining the endpoint to call, prepare the data to send with the request if needed, submit the request, and, finally, receive and process the response.

.NET has a built-in class called `HttpClient`, which is a rich class that provides all the functionality and configuration needed to send any type of request (GET, POST, PUT, PATCH, DELETE, and more) to a given web API.

Understanding FetchData component logic

All Blazor WebAssembly projects have a default page called FetchData in the Pages folder. The logic of this page is to make a GET request to download the content of a JSON file that lives in the wwwroot folder, and that content also represents weather forecast data. So, let's analyze the code of the FetchData component so that we understand it better:

```
@page "/fetchdata"
@inject HttpClient Http
...
            @foreach (var forecast in forecasts)
            {
                <tr>
                    <td>
                        @forecast.Date.ToShortDateString()
                    </td>
                    <td>@forecast.TemperatureC</td>
                    <td>@forecast.TemperatureF</td>
                    <td>@forecast.Summary</td>
                </tr>
            }
...

@code {
    private WeatherForecast[]? forecasts;
    protected override async Task OnInitializedAsync()
    {
        forecasts = await
          Http.GetFromJsonAsync<WeatherForecast[]>("sample-
          data/weather.json");
    }

    public class WeatherForecast
    {
        public DateTime Date { get; set; }

        public int TemperatureC { get; set; }

        public string? Summary { get; set; }

        public int TemperatureF =>
          32 + (int)(TemperatureC / 0.5556);
    }
}
```

In the previous code snippet, we notice the following:

- In the second line in the snippet, there is an instance of the `HttpClient` type called `Http` injected into the component.

- In the `@code` section, there is a class called `WeatherForecast`, and the properties inside represent the data in the weather forecast JSON object, which we will see in *step 5*.

- There is a declaration for a `WeatherForecast` array called `forecasts`.

- The most important part for us is the line inside `OnInitializedAsync`, which is basically the GET call to download the content of the file. The code provided uses the injected `Http` object and the `GetFromJsonAsync` method; this is an extension method for the `HttpClient` class, and it sends a GET request to a specific URL (`sample-data/weather.json` in this example), reads the JSON content that comes with the response, and deserializes it to a defined type – in this case, it's deserialized to an array of `WeatherForecast`. Finally, it assigns that value to the `forecasts` variable.

  We will talk more about the origin of the base URL for the request after this section.

- The request refers to a JSON file in the `sample-data` folder called `weather.json`. We can find the folder and the file in the `wwwroot` folder of the project, and if we open it, we will see the following JSON:

  ```
  [
    {
      "date": "2018-05-06",
      "temperatureC": 1,
      "summary": "Freezing"
    },
    ...
  ]
  ```

  JSON represents an array of an object that contains three properties – `date`, `temperatureC`, and `summary`. The `WeatherForecast` class has been defined based on these properties.

- Finally, in the markup section, there is an `if` statement to check whether the `forecasts` array is `null`, and if not, it creates an HTML table, iterates over each item in the `forecasts` array, and renders a row for it in that table.

The preceding is an explanation of the default code that uses `HttpClient` to get content from a server. Before we write our first call to fetch the books from the `BooksStore` API, instead of getting them from an in-memory list, let's see where that `HttpClient` instance comes from and how it's configured.

Configuring HttpClient in Blazor WebAssembly

Because the `FetchData` component injected an instance of `HttpClient`, the object is already registered in the dependency injection container. By default, Blazor did that for us. Let's open the `Program.cs` file, and we will find the following line:

```
...
builder.Services.AddScoped(sp => new HttpClient { BaseAddress = new
Uri(builder.HostEnvironment.BaseAddress) });
...
```

This registers a new object of the `HttpClient` type configured with a base address via the `BaseAddress` property. This property sets the base URL of every call you send. Because, by default, it is set to the root URL of the application, the `GetFromJsonAsync` method in the `FetchData` component directly points to `/sample-data/weather.json` without the need to set the full URL of the app.

The `BooksStore` API has a different URL. During development, it will be `https://localhost:7188`, and when we publish it, it will be something else, so we need to modify the `BaseAddress` property in the current `HttpClient` registration.

So, let's get that done first:

1. Open the wwwroot folder and the `appsettings.Development.json` file we created in *Chapter 1, Understanding the Anatomy of a Blazor WebAssembly Project*, in which we talked about environments. We added a property called `ApiUrl`, so let's modify its value to the URL of the web API so that the JSON file will look like this:

    ```
    {
        "ApiUrl": "https://localhost:7188"
    }
    ```

    Setting the URL here instead of hardcoding it into `Program.cs` will make it easier to develop, and when we push to production, we can set the production URL of the API in `appsettings.json`, which will be used by the app when it's in the production environment. So, there's no need to change the URL in `Program.cs` during development and then change it later before publishing.

2. Turning back to `Program.cs`, change the value of the `BaseAddress` property to make it refer to the `ApiUrl` value from the configuration, as shown here:

    ```
    ...
    builder.Services.AddScoped(sp => new HttpClient { BaseAddress =
    new Uri(builder.Configuration["ApiUrl"]) });
    ...
    ```

Now, we are ready to start using this `HttpClient` instance to send API requests to our `BooksStore` API. However, by changing the base URL, the current `FetchData` component will stop working because its GET request refers to the JSON file in the base URL of the Blazor WebAssembly project. We need to fix this.

3. In the `FetchData.razor` component in the `Pages` folder, we can either point to the `/WeatherForecast` GET endpoint that the API offers, or we will need to set the full URL of the application in `GetFromJsonAsync` so that it overrides the URL of the API we set in `Program.cs`. The first solution is better, as it keeps the API source of our project unified, so let's update that line as follows:

```
. . .
protected override async Task OnInitializedAsync()
{
    forecasts = await
      Http.GetFromJsonAsync<WeatherForecast[]>(
        "/WeatherForecast");
}
. . .
```

That will fetch the same data from the `/WeatherForecast` endpoint we tested in the previous section with Postman.

To make sure everything is fine, let's run the project and navigate to `/FetchData` and make sure the web API is running too. You should be able to see the table of the weather data populated normally, but this time, the data will come from the web API, not the JSON file in the `wwwroot` folder.

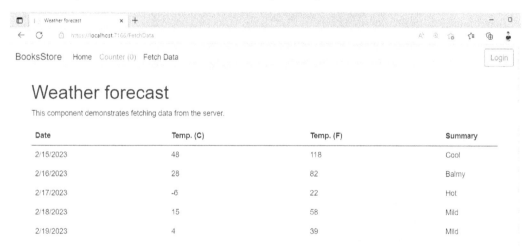

Figure 8.7 – Updated FetchData page to get the weather data from the API

Sending a GET request

Let's learn more about calling web APIs and improve our project even further. When you run the project, it redirects you by default to the index page, where a list of books will be shown on the UI. The books are fetched from a local in-memory collection inside the `LocalBooksService.cs` class within the `Services` folder. We need to replace that fixed data list with a web API call that retrieves the books from the API. Unlike the `GET` request we saw in the `FetchData` component, this one will be written step by step, and we will have better control over the response:

1. As we learned earlier, before we write the code, we need to understand the targeted endpoint. Navigate to the web API Swagger page and expand the `/Books` GET request to see what it returns in both success and failure cases.

2. You can open Postman and send a GET request to `https://localhost:7188/books` to see the response it retrieves.

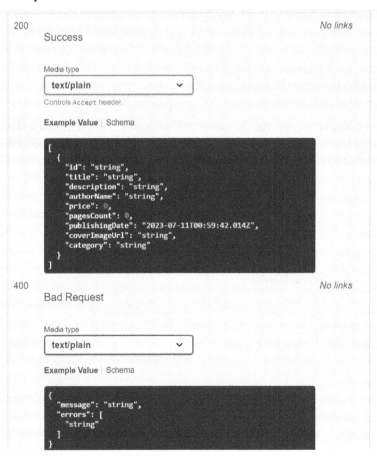

Figure 8.8 – GET /Books request responses in web API

The documentation says that the endpoint either returns the *status code 200*, which indicates success, or *400*, which indicates failure. In a successful case, the content of the response will be an array of books, and, in the case of failure, it will return an object that contains a message property of the string type and an array of strings called errors. We deserialize these two objects based on the desired status into corresponding C# classes.

3. Earlier, we created a `Book` class and the `coverImageUrl` property coming from the API that we need to add to the model. You also need to make sure that the JSON properties coming from the API are mapped correctly to the C# object you have. So, we will use the `[JsonPropertyName()]` attribute for each C# property in the `Book` class. In the current case, the names of the JSON properties are the same for our `Book` class, but that won't always be the case, so it's a good practice to map them explicitly, as shown here:

```
public class Book
{
    [JsonPropertyName("title")]
    public string? Title { get; set; }
    [JsonPropertyName("authorName")]
    public string? AuthorName { get; set; }
    [JsonPropertyName("publishingDate")]
    public DateTime PublishingDate { get; set; }
    [JsonPropertyName("price")]
    public decimal Price { get; set; }
    [JsonPropertyName("description")]
    public string? Description { get; set; }
    [JsonPropertyName("coverImageUrl")]
     public string? CoverImageUrl { get; set; }
    [JsonPropertyName("id")]
    public string? Id { get; set; }
    [JsonPropertyName("pagesCount")]
    public int PagesCount { get; set; }
}
```

4. Create a C# class that corresponds to the failure object based on the documentation in the `Models` folder and call it `ApiErrorResponse`:

```
public class ApiErrorResponse
{
    [JsonPropertyName("message")]
    public string? Message { get; set; }
    [JsonPropertyName("errors")]
    public string[]? Errors { get; set; }
}
```

5. Open the `Index.razor` page in the `Pages/UserPages` folder, and let's inject an instance of the `HttpClient` class and call it `Http`:

    ```
    . . .
    @inject HttpClient Http
    . . .
    ```

6. In the `@code` section, create a method called `GetBooksAsync` that has the following code:

    ```
    . . .
    private async Task<List<Book>?> GetBooksAsync()
    {
        var response = await Http.GetAsync("/books");
        if (response.IsSuccessStatusCode)
        // if response status code is 2XX
        {
            return await response.Content
                .ReadFromJsonAsync<List<Book>>();
        }
        else
        {
            var errorResponse = await response.Content
                .ReadFromJsonAsync<ApiErrorResponse>();
                // Throw an exception with the error
                // message for now
            throw new Exception(errorResponse?.Message);
                // TODO: Handle the error in Chapter 11
        }
    }
    . . .
    ```

Let's explain what this method does step by step. First, it uses the `Http.GetAsync` method to send a GET request to the `/books` endpoint in the web API. We refrained from using the `GetFromJsonAsync` extension method in this scenario because it does not handle unsuccessful responses effectively. If the response does not indicate success, `GetFromJsonAsync` throws an exception, making it difficult to control the error-handling process. `GetAsync` will retrieve an object of the `HttpResponseMessage` type, and we can handle successful and failed cases manually.

After that, we check whether `IsSuccessStatusCode` is set to true in the response object. This property is set to true if the response status code is 2XX. If the request succeeds, the endpoint documentation tells us that it returns a collection of `Book` objects, so we need to read the JSON content of the response and deserialize it into a list of books. The `Content` property within the response object has a set of methods that allow us to read the content in different ways, and there is an extension method called `ReadFromJsonAsync` that reads the content of the response and deserializes it into a specific type. So, we read and deserialize the content to `List<Book>` and return it.

If the request failed, we know that the API will return an `ApiErrorResponse` object, so we deserialize the content to that type and, for the time being, we will throw an exception, but we will look at error handling in more depth in *Chapter 10, Handling Errors in Blazor WebAssembly*.

7. Call the `GetBooksAsync` method within `OnInitializedAsync` and assign the returning value to the existing `_books` variable that's being rendered in the UI. We can achieve that by commenting out the current code that fetches from `BooksService` and adding the call we just created, as follows:

```
protected async override Task OnInitializedAsync()
{
...
    //_books = await BooksService.GetAllBooksAsync();
    _books = await GetBooksAsync();
}
```

That's how you can make a GET request in detail, from understanding the request and the response it returns to creating the model, sending the request, and finally running the app alongside the API and navigating to the **Index** page. The app should list all the books retrieved from the web API.

> **Note**
>
> Sometimes, you may not be interested in all the properties returned by the API, so you can add only what you need to the corresponding C# models you are creating.

Now that we have learned how to fetch data from the API, it's time to learn how to send data to it by implementing a POST request in the `BookForm` component. The request should hold book information for the API so that the API can save the book in the API data store.

Implementing a POST web API call in Blazor WebAssembly

We have developed the `BookForm` page, which has a form for entering the book details and a submit button that currently does nothing other than write some logs to the console window.

The `BookForm` page is for the store admin, who can insert new books. For now, we don't have authentication either on the server side in the web API or on the client side; that's the target of *Chapter 9*. But we need to complete the logic of this page so that when the admin clicks the **Submit** button, the book gets posted to the API, which stores it in its data source. When the app goes online in production and an admin adds a new book, all those users who navigate to the app will see the new book on the **Index** page because that web API is the main source of data for our project.

The `BooksStore` API exposes a POST endpoint, and according to the Swagger documentation, accepts a JSON object in the body of the POST request, as shown here:

```
{
    "title": "string",
    "description": "string",
    "author": "string",
    "price": 0,
    "pagesCount": 0,
    "publishingDate": "2023-02-14T02:11:26.158Z"
}
```

In the `Models` folder, we have a model called `SubmitBook` that's used by the **BookForm** page. We need to match the properties in the JSON of the POST request to the properties' names in the `SubmitBook.cs` file. We will do that using `[JsonPropertyName()]`, as shown here:

```
public class SubmitBook
{
    [Required]
    [StringLength(80, MinimumLength = 3)]
    [JsonPropertyName("title")]
    public string? Title { get; set; }

    [StringLength(5000)]
    [JsonPropertyName("description")]
    public string? Description { get; set; }
    [Required]
    [StringLength(80, MinimumLength = 3)]
    [JsonPropertyName("authorName")]
    public string? Author { get; set; }

    [Range(typeof(decimal), "0", "99999")]
    [JsonPropertyName("price")]
    public decimal Price { get; set; }

    [DisplayName("Number of Pages")]
    [Range(0, 9999, ErrorMessage = "Number of pages must be
      at least 0 and at most 9999")]
    [JsonPropertyName("pagesCount")]
    public int PagesCount { get; set; }
}
```

Now, we need to write the send POST request logic in the `BookForm.razor` component inside the `Pages` folder:

1. Inject an instance of `HttpClient` at the top of the component and call it `Http`, and inject an instance of `NavigationManager` to navigate the user after the book is submitted successfully:

   ```
   . . .
   @inject HttpClient Http
   @inject NavigationManager NavigatinManager
   . . .
   ```

2. In the `@code` section, update the `HandleBookFormSubmission` method by removing the `Console.WriteLine()` calls and add the submit POST request logic as follows:

   ```
   . . .
   private async void HandleBookFormSubmission()
   {
       _book.Description =
         await _simpleMde.GetEditorValueAsync();

       var response = await Http.PostAsJsonAsync("books",
                                                 _book);
       if (response.IsSuccessStatusCode)
       {
           await ClearSavedStateAsync();
           NavigationManager.NavigateTo("/");
       }
       else
       {
           var error = await response.Content
             .ReadFromJsonAsync<ApiErrorResponse>();
           Console.WriteLine(error);
           // TODO: Handle the error
       }
   }
   . . .
   ```

Let's explain the changes. First, we removed all the logging calls to the console window; then, we used the `HttpClient` instance to send a POST request by calling `PostAsJsonAsync`. Because in the POST request, we are sending data to the web API, `PostAsJsonAsync` takes two parameters: one is the endpoint we want to reach, and the other is the object we want to send as JSON. The method will serialize the object to JSON and send the request.

We store the response in the response object. Then, as we did earlier, we validate the status of the response. If the response succeeded, that means the book has been added successfully. We then remove the saved objects to preserve the state using the `ClearSavedStateAsync` method and navigate the user to the **Index** page.

If the request fails, we basically deserialize the content of the response into an `ApiErrorResponse` object, and for now, just log the error on the console because we will learn how to handle errors efficiently in *Chapter 10*.

Now, if you run the project with the API and navigate to `/BookForm`, populate the data, and hit **Submit**, you will be redirected to the **Index** page and see the new book in the list.

With that, we have learned what's needed to communicate with a web API using the `HttpClient` class. Next, we will explore the `IhttpClientFactory` service briefly, in addition to covering the delegating handlers, what they are, and when you need them.

Now that we have made the calls and exposed our app to real data, let's see how we can get that code clean so the components stay readable, small, and testable, as we will see in *Chapter 13, Testing Blazor WebAssembly Apps.*

Exploring IHttpClientFactory and delegating handlers

`IHttpClientFactory` is a service that supports creating single or multiple `HttpClient` instances with custom configuration. We have used the `HttpClient` instance configured in the previous section with the web API base URL. This was enough to meet our basic needs, but sometimes slightly more advanced techniques are needed to cover the scenarios you may face while developing more complex apps.

In this section, we will learn how to use the `HttpClient` factory to create and manage `HttpClient` instances. In addition, we will learn about `DelegatingHandlers` that allow for building a pipeline to process each HTTP request or response associated with a specific `HttpClient` instance and when we need such a capability.

To be able to utilize the `IHttpClientFactory` service, we need to install the `Microsoft.Extensions.Http` package. You can do that either using the NuGet package manager or through the .NET CLI command:

```
dotnet add package Microsoft.Extensions.Http
```

The first concept we will introduce is the named `HttpClient` instances. Currently, we have a new instance of `HttpClient` registered in the DI container. Using `IHttpClientFactory`, we can register one or more `HttpClient` instances with a name for each. Then, we can retrieve any of those `HttpClient` instances using the name we registered it with. So, in `Program.cs`, we will replace the line that manually registers `HttpClient` with the `AddHttpClient` method and give it the name `BooksStore.Api`, as follows:

```
...
//builder.Services.AddScoped(sp => new HttpClient { BaseAddress = new
Uri(builder.Configuration["ApiUrl"]) });
builder.Services.AddHttpClient("BooksStore.API", httpClient =>
httpClient.BaseAddress = new Uri(builder.Configuration["ApiUrl"]));
...
```

As a result of the previous code, there are no `HttpClient` instances registered directly in the DI container. So, to be able to retrieve this instance of `HttpClient` you need to inject the `IHttpClientFactory` service, then use the `CreateClient` method, as shown in the following snippet:

```
@inject IHttpClientFactory HttpClientFactory
...
using (var httpClient =
        HttpClientFactory.CreateClient("BooksStore.API"))
{
    // Use the httpClient here
}
```

This approach does require us to make some code changes wherever we inject an `HttpClient` instance. But because we only have one instance of `HttpClient` to use throughout the full project, we can retrieve this instance from the `IHttpClientFactory` service and register it in the DI container directly by adding this line to the `Program.cs` file:

```
...
builder.Services.AddScoped(sp =>
sp.GetRequiredService<IHttpClientFactory>().CreateClient("BooksStore.
API"));
...
```

The preceding line retrieves `IHttpClientFactory` and uses the `CreateClient` method to create the required `HttpClient` instance and then registers this in the DI container. With this implementation, no code changes are required for what we have accomplished in the previous sections. Every time you inject `HttpClient`, instead of creating a new instance in the way it did before, a new instance will now be created using the factory.

We have now reached the same point we were at before this section, but the difference is that now we will add some features to that `HttpClient` instance, such as `DelegatingHandler`.

What are delegating handlers? And why do we need them? As a simple definition, you can think of a delegating handler as a gateway that every HTTP request and response for a specific `HttpClient` instance will go through. This allows for a centralized place to manipulate the HTTP request or the response based on your needs. You can have one or a chain of delegating handlers associated with a single `HttpClient` so that each request or response goes from one handler to the next, where each delegating handler is responsible for a certain manipulation task.

There are many scenarios where delegating handlers can be useful. Let's take logging as an example. Let's say you want to log each HTTP request and response – instead of writing these logs manually in each method where you send a request, you can build a delegating handler and write the logs there so every HTTP request will be logged.

In *Chapter 9*, we will learn about protected API endpoints, covering how, to be able to access them, we need to send an access token in the header of each HTTP request. In this scenario, a delegating handler is needed to centralize this process for all the requests. We will go over the authentication scenario in the next chapter in detail. To keep things simple for now and learn how to build a delegating handler, we will create a simple one that logs the URL of each request and the status code for each response; then, we will link that basic handler to the existing `HttpClient` instance that we have.

To create this example logging delegating handler, create a new class in the root folder of the project with the name `DemoLoggingHandler.cs`, then add the following piece of code:

```
public class DemoLoggingHandler : DelegatingHandler
{
    private readonly ILogger<DemoLoggingHandler> _logger;

    public DemoLoggingHandler(ILogger<DemoLoggingHandler>
                                logger)
    {
        _logger = logger;
    }

    protected override async Task<HttpResponseMessage>
      SendAsync(HttpRequestMessage request,
                CancellationToken cancellationToken)
    {
        _logger.LogInformation($"HTTP request sent
          {request.Method} {request.RequestUri}");
        var response = await base.SendAsync(request,
          cancellationToken);
        _logger.LogInformation($"HTTP response received
                                :{response.StatusCode}");
        return response;
    }
}
```

Building handlers is easy and straightforward: just inherit from the `DelegatingHandler` class and override the `SendAsync` method that supplies the request as a parameter of type `HttpRequestMessage`. You can use this parameter to manipulate the request or the response as we did in the code after sending the request using the `SendAsync` method.

The last thing we need to do is to attach `DemoLoggingHandler` to the `HttpClient` instance. In `Program.cs`, register an instance of `DemoLoggingHandler` to the DI container and use the `AddHttpMessageHandler<DemoLoggingHandler>` extension method to the `AddHttpClient` method as follows:

```
...
builder.Services.AddScoped<DemoLoggingHandler>();
builder.Services.AddHttpClient("BooksStore.API", httpClient =>
httpClient.BaseAddress = new Uri(builder.Configuration["ApiUrl"]))
                .AddHttpMessageHandler<DemoLoggingHandler>();
...
```

That's all. Easy, right? Now run the project. In the index page, when the API request is sent to retrieve the books, and for every future API request we will add, you will automatically notice the following logs in the **Console** window of the browser dev tools:

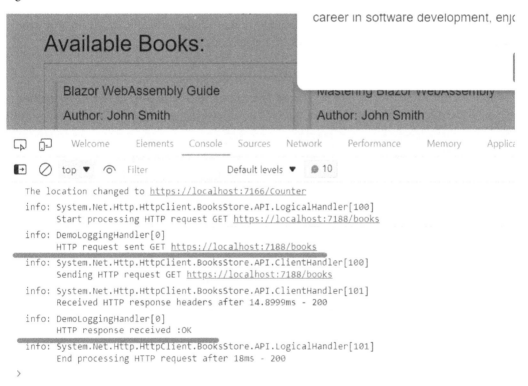

Figure 8.9 – Logs from the delegating handler

We will build more on top of this in the next chapter, but for now, we have learned everything we need to get our Blazor WebAssembly app to communicate with web APIs. The last thing we will achieve in the next section is to take the code to send the API requests from within the components into a separate service, thus ensuring the readability and testability of the code.

Separating your API calls from the components

Back in the *Dependency Injection in Blazor WebAssembly* section in *Chapter 1*, we created the IBooksService interface and an implementation for it called LocalBooksService. The goal was to make the components rely on the IBooksService interface and not the implementation, which can be changed. The code to call the web API endpoints is somewhat long and repeatable, in addition to making the components depend on an instance of HttpClient, which makes the components hard to test and have longer code.

The goal in this section is to migrate the calls we made in the previous two sections to IBooksService and create a new implementation for it that communicates with the web API. Finally, we will write some code to fetch some book details from the API using book IDs.

So, let's get started:

1. In the IBooksService interface in the Services folder, we need to add a new method called AddBookAsync that takes a SubmitBook object as a parameter. The method won't return anything – it either passes and the book is added successfully, or it fails and it will throw an exception:

    ```
    public interface IBooksService
    {
        Task<List<Book>> GetAllBooksAsync();
        Task<Book?> GetBookByIdAsync(string? id);
        Task AddBookAsync(SubmitBook book);
    }
    ```

2. In the Services folder, create a new C# class and call BooksHttpClientService. cs, which will be an implementation for IBooksService but communicate with the API instead of a local list.

3. Inside BooksHttpClientService.cs, implement the IBooksService interface and inject an instance of type HttpClient:

    ```
    using BooksStore.Models;
    using System.Net.Http.Json;

    namespace BooksStore.Services;
    public class BooksHttpClientService : IBooksService
    {
    ```

```
        private readonly HttpClient _httpClient;
        public BooksHttpClientService(HttpClient
            httpClient)
        {
            _httpClient = httpClient;
        }
    }
```

4. Implement the `GetAllBooksAsync` method, copy the code written in the **Index** page to call the web API, and put it inside the method, as shown here:

```
    ...
        public async Task<List<Book>?> GetAllBooksAsync()
        {
            var response =
                await _httpClient.GetAsync("books");
            if (response.IsSuccessStatusCode)
            // if response status code is 2XX
            {
                return await response.Content
                    .ReadFromJsonAsync<List<Book>>();
            }
            else
            {
                var errorResponse = await response.Content
                    .ReadFromJsonAsync<ApiErrorResponse>();
                // Throw an exception with the error
                // message for now
                throw new
                    Exception(errorResponse?.Message);
            }
        }
    ...
```

5. The API exposes a *get book by id* endpoint, which is a `GET` request and is accessible through the `/books/{Book_Id}` route. If the request succeeds, it returns the same `Book` object that the `/books` endpoint returns. So, we can replicate the call but change the return type to a single book and add the ID to the route as follows:

```
        public async Task<Book?> GetBookByIdAsync(string?
                                                    id)
        {
            var response = await
                _httpClient.GetAsync($"books/{id}");
```

```
        if (response.IsSuccessStatusCode)
        // if response status code is 2XX
        {
            return await
            response.Content.ReadFromJsonAsync<Book>();
        }
        else
        {
            var errorResponse = await
                response.Content.ReadFromJsonAsync
                <ApiErrorResponse>();
            throw new
                Exception(errorResponse?.Message);
        }
    }
```

6. We still have to implement AddBookAsync, so we need to bring the HTTP call we made in BookForm and paste it inside the form, but make sure to remove the component-related code, such as navigation and clearing the state, because in this service we just need the code to call the API:

```
public async Task AddBookAsync(SubmitBook book)
{
    var response = await
        _httpClient.PostAsJsonAsync("books", book);
    if (!response.IsSuccessStatusCode)
    {
        var error = await
            response.Content.ReadFromJsonAsync
            <ApiErrorResponse>();
        Console.WriteLine(error);
    }
}
```

After sending the request, if it succeeds, then no action is needed in the service here; we just need to handle the failure.

7. Because we have a new implementation for IBooksService, we can register the dependency injection container to use an instance of the BooksHttpClientService instead of LocalBooksService.

8. In Program.cs, comment out the existing line and add a new one, as shown here:

```
...
//builder.Services.AddScoped<IBooksService,
LocalBooksService>();
```

```
builder.Services.AddScoped<IBooksService,
BooksHttpClientService>();
. . .
```

9. The service is ready, and to use the web API calls, we just need to inject an instance of
    IBooksService. Now we need to clean the calls we made earlier.

    In Index.razor, remove the injection of the HttpClient instance, remove the
    GetBooksAsync method, and change the call inside OnInitializeAsync to be as follows:

    ```
    protected async override Task OnInitializedAsync()
    {
    . . .
        _books = await BooksService.GetAllBooksAsync();
        //_books = await GetBooksAsync();
    }
    ```

    The preceding code is back to its original state. We worked through this whole exercise so you
    could get a deeper understanding of the service separation and remain focused on making the
    API calls at the initial state, then worry about making it clean later.

10. The same thing needs to be repeated in the BookForm component. We will remove the
    HttpClient injection and inject IBooksService instead:

    ```
    . . .
    @inject IBooksService BooksService
    . . .
    ```

11. Refactor the HandleBookFormSubmission method as follows:

    ```
    private async void HandleBookFormSubmission()
    {
        _book.Description =
          await _simpleMde.GetEditorValueAsync();
        await BooksService.AddBookAsync(_book);
        await ClearSavedStateAsync();
        NavigationManager.NavigateTo("/");
    }
    ```

    You can see here that there is no error handling and no data is returned, and that's because the
    AddBookAsync method either adds the book successfully and doesn't return anything, or
    fails and throws an exception, which we will handle in *Chapter 11*.

The code difference between having the API calls within the components and separating them into
external services is significant. When the calls are separated, there are no long, repeated API calls
inside the components and they don't depend on an instance of HttpClient anymore. However, the

components we have refactored now depend on an interface of the `IBooksService` type. *Chapter 13, Testing Blazor WebAssembly Apps*, will detail how this makes our component much more testable.

We have come a long way, right? Our application is now more interactive and has better code, and there is less fake or simulation logic. We have leveled up by starting to think about how the code looks and how long it is, and trying to make it as small as possible so that instead of having code that just works, we get code that works efficiently and looks great too.

Summary

We started this chapter by explaining the concept of web APIs regardless of your level of expertise so that you could get a deeper understanding of what they are and why you need them. Then, we moved on to the practical side of things by introducing the tools and utilities available to make API calls from a .NET app. We implemented two API calls: one to fetch data from the API and another to submit data to the API. Finally, we went over what we did, refactored it, and learned about the correct place for the API calls to reside so we have more testable code and better-coded components.

After going through the content and examples in this chapter, you should be able to consume your web APIs on your own and integrate your apps with third-party APIs to enrich the experience of your users.

The following is what you should have gained from this chapter:

- The ability to identify what web APIs are from the client perspective and why they are needed
- Read and understood the documentation of APIs, identified the available endpoints, and now know how to analyze and test them
- Learned how to make web API calls efficiently from your Blazor WebAssembly project
- Explored and used `IHttpClientFactory` alongside `DelegatingHandler`
- Learned how to separate your calls into different services for better code readability and testability

Now that we have exposed the app to an online service, in the next chapter, we are going to get some users into the system and allow them to create new accounts and log in. We also need to protect some pages and restrict access to some functionality to the admin only. In the upcoming chapter, we will deep dive into how to secure our Blazor WebAssembly app.

Further reading

- *Call APIs from Blazor WebAssembly*: `https://learn.microsoft.com/en-us/aspnet/core/blazor/call-web-api?view=aspnetcore-7.0&pivots=webassembly`

9

Authenticating and Authorizing Users in Blazor

All the features we have built and learned about so far are for any user who loads the app in the browser. Well, in real-world applications, that's not always the case. In most of the scenarios you will face, some features require the app to know who is using it, and in other cases, the full app cannot be used until it knows who is using it.

In this chapter, we are going to learn about authentication and how it happens on the client side, what authentication means, and how we can implement it in Blazor WebAssembly.

We will explore the Blazor authentication library in depth, alongside most of its capabilities that enable you to build enterprise-level single-page applications. To understand everything in detail, we will build a custom authentication flow. After building the core, we will take advantage of the authentication library to restrict access to some pages to only logged-in users, show or hide partial content in the UI, and control the logic of a component based on the user's identity. After that, we will implement role-based access control to restrict some features in the app for specific roles. If this is not enough for our use case, we will go further and discuss authorization policies and put them into practice.

Another point to discuss is calling secured API endpoints from the client by sending the token with each HTTP request. Finally, we will take a look at the **Microsoft Identity Platform** and services such as **Azure Active Directory** and **Active Directory Business-to-Customer** (**B2C**) and what benefits they will give us if we use them with Blazor.

The following topics will be covered in this chapter:

- Understanding authentication in Blazor WebAssembly
- Implementing a custom authentication flow
- Implementing authorization and advanced authentication features

- Accessing authorized API endpoints
- Securing a Blazor WebAssembly app with Azure **Active Directory (AD)**

Technical requirements

The Blazor WebAssembly code used throughout this chapter is available in the book's GitHub repository in the *Chapter 9* folder: `https://github.com/PacktPublishing/Mastering-Blazor-WebAssembly/tree/main/Chapter_09/Chapter_Content`.

The API project we used in *Chapter 8* will also be needed, as more endpoints have been added to it. You can find the project in the API folder in the GitHub repository folder for *Chapter 9*: `https://github.com/PacktPublishing/Mastering-Blazor-WebAssembly/tree/main/Chapter_09/API`.

In the last section, we will introduce Azure AD and AD B2C. If you are curious about using them with Blazor WebAssembly, an Azure account with an active subscription is required. If you don't have an Azure account already, you can get started for free at the following link: `https://azure.microsoft.com/en-us/free/`.

Understanding authentication in Blazor WebAssembly

With the word authentication, probably the first thing that comes to your mind is the login process with a username and password. I think that's mostly right, but the story is a bit deeper than that.

An overall general definition for **authentication** is validating whether the user or the system that is trying to access certain protected resources has a valid identity or not. Think of authentication as an employee trying to access a company office. The employee passes the work ID over the scanner beside the main entrance. The system then checks whether it is a valid ID and, if it is, the door opens and allows the employee in. The employee gets their work ID in the onboarding process after getting hired, so you can map the entrance validation to the login process and the issuing of the ID to the registration process in the software world.

Authentication is not only for users (actual people who open and use the app). It can also be against another system. When two systems are trying to communicate with each other and system A is accessing protected data or executing protected actions with system B, then system B has to authenticate system A to make sure it's the actual system that has the right to access the requested resources. All these authentication processes happen in the backend and mostly between APIs, which is a bit beyond the scope of this book.

It's good to differentiate between authentication on the client side and the server side. On the client side, authentication doesn't guarantee security. The actual security processes of the system occur on the server side when we ask the API to execute an action or return some data. At that moment, the API checks whether we are eligible to execute the action or are the rightful owners of this data. On the client side, especially on apps that run in the browser, authentication doesn't strictly mean security; rather, it means showing and hiding UI elements based on the user's identity. The stuff being controlled is just pure UI, such as not showing a page, a link in the nav menu, or a button that executes an action to non-admin users. However, always keep in mind that client-side authentication can be bypassed, so the actual security of the data and the system is on the server.

In relation to single-page applications, which we are focusing on in this book, our authentication target is the person who is using our app. In the case of the `BooksStore` app that we are building, not all the features require an authenticated user. For example, if someone opens the URL of our app and browses the books, that should be fine and we don't need to know who that user is exactly, but if the user wants to buy a book or write a book review, then we need to know who this person is.

Another part of the system is admin. Certain employees of our books library should have admin access to the system so that they can manage the books (add, edit, and delete), so to access this part of the system, they must be authenticated. So, we can see that our system has some pages, such as the book details page, that can be accessed without authentication, but some features inside it must be blocked, such as writing a review. Other pages, such as the page for adding a new book, cannot be accessed at all without a user being logged in.

So, how can we get this stuff done? Maybe that's the question you have in mind right now. A naïve solution will be as follows:

1. Create a login page that includes email and password fields and a submit button.
2. Send a `POST` request to the API with the email and password, and it should retrieve a user object if it finds one.
3. Store the user object somewhere in a static class called `UserState`, for example.
4. Then, the user is authenticated if the `user` object inside the `UserState` class is not `null`.

That seems cool, but the actual process is not 100% straightforward in the way that the preceding list suggests. Fortunately, Blazor provides you with all the infrastructure needed to enable authentication in your app efficiently.

Next, we will discover what Blazor offers us out of the box to build an authentication system for our single-page application, and how to utilize those utilities.

It's good to know that Blazor WebAssembly runs fully in the browser, so authentication will mostly mean showing or hiding a functionality or page to or from a user based on certain conditions, which we will see later, but keep in mind that all client-side in-browser authorization can be bypassed in one way or another. So, your API security can enforce the security rules needed.

Token-based authentication (TBA)

In Blazor WebAssembly and almost every other single-page application framework, it's common to use what's called **TBA**. TBA is simply an authentication method that uses a token fetched from an identity provider (your web API or an external identity provider such as Azure Active Directory) after the user logs in successfully. This token is an encoded JSON string that represents some data about the user, in addition to a signature confirming the validity of the token; this token is also known as a **JSON Web Token (JWT)**. The Blazor WebAssembly app gets the token, parses it, extracts the user data, which is known as **claims**, constructs a user object out of those claims, then stores the token in the protected browser storage.

The following screenshot shows a sample JWT, decoded using the `https://jwt.ms` online utility.

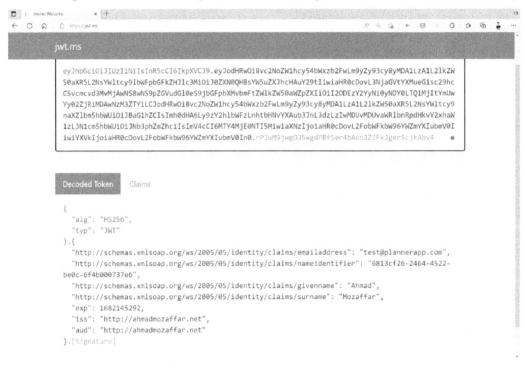

Figure 9.1 – JWT decoded

As you can see in the preceding screenshot, the first box contains the raw token, which consists of three main parts:

- The first part, which is highlighted in red, represents the type of token and the hashing algorithm it uses.

- The second long part, highlighted in blue, represents the token payload or the data it holds that the identity server has exchanged with the Blazor app.

- The third part, highlighted in green, represents the signature of the token. Basically, this is the previous parts combined and encrypted with a specific key.

Decoding the token and extracting the claims from the payload and building a user object out of them is what enables the Blazor WebAssembly app to mark a user as authenticated. Then, the token can be stored in the protected browser storage and used in each HTTP request sent to the API to access the protected endpoints.

When we use the term "user claims," we are referring to a key-value pair that defines the user. In *Figure 9.1*, in the **Decoded Token** section, you can see a JSON object highlighted in blue. This object contains a set of properties along with their values, which are called claims. Some examples of claims include email addresses, name identifiers (IDs), given names, and surnames. There are also special claims, such as `exp`, which represents the expiry date of the token.

Now that we've introduced TBA, let's see what Blazor provides to achieve the aforementioned logic in the code.

The Blazor authentication library

Blazor provides the **Microsoft.AspNetCore.Components.WebAssembly.Authentication** package, which wraps all the required utilities to support authentication and authorization in your apps. The package contains classes and services that support different scenarios and authentication protocols, such as **OpenID Connect (OIDC)**. OIDC is a big topic to explain, but in a nutshell, it's the protocol you see when you use any app that allows you to authenticate using your Facebook or Google account. When you click **Login with Google**, for example, the app redirects you to Google and you log in with your Google account, then it directs you back to the app and you are logged in successfully. OIDC is also based on JWT, so the authentication will be achieved as mentioned in the previous chapter.

Alongside the aforementioned library, Blazor WebAssembly uses some capabilities from the ASP. NET Core authentication library, **Microsoft.AspNetCore.Authorization**. Both libraries provide all the ingredients to support all authentication scenarios, and here are some of the classes, attributes, and components the packages contain that help achieve the authentication mechanism:

- **AuthenticationStateProvider class**: This is the core of the authentication process. It provides a method called `GetAuthenticationStateAsync`, which is used to build an `AuthenticationState` object containing a property called `User` of the `ClaimsPrincipal` type, and this is what wraps the claims of the user. It also has an event that gets triggered when the authentication state changes in the app.

- **[Authorize] attribute**: This is used to secure pages. When we declare a page with `[Authorize]`, if the user is not logged in, they won't be able to access this page.

- **AuthorizeRouteView component**: In the *Building layouts in Blazor* section in *Chapter 3*, we talked about a default component called `RouteView` used in the `App.razor` component. When your app is using authentication, you should consider using `AuthorizeRouteView` instead, as this will check whether the requested page requires authentication or not.

- **AuthorizedView component**: This component helps us render specific content based on the user's authentication status.

- **CascadingAuthenticationState component**: This can be used to wrap all other components in the `App.razor` component. It provides the `Task<AuthenticationState>` object as a cascading parameter to any ascending component that wants to access the current user identity state and the user claims.

Throughout this book, we have started from the very basic components and kept going until we have reached the target and built the advanced component we are looking for. The same process will be followed now, so after introducing the theoretical part, let's get started putting all those components together to make our app secure.

Building a custom JWT authentication flow

This is the section that will contain all the action! We will put together everything we have learned so far about developing components and forms and calling a web API, in addition to all the parts we mentioned in the previous section, to add authentication to our `BooksStore` project.

For this exercise, we will use the `/authentication/login` POST API endpoint. This endpoint will accept an object with two properties – `Username` and `Password`. If they are valid, it will return an object containing the access token.

By default, the API has two users registered that we can use to test:

- *John Smith*: He is an admin in the company. His email is `admin@masteringblazor.com` and his password is `Test.123`.

- *Ahmad Mozaffar*: He is a customer of the `BooksStore` library. His email is `ahmad.mozaffar@masteringblazor.com` and his password is `Test.123`.

The custom flow we will build will consider the user authenticated when there is a valid token stored in the local storage of the browser.

Let's summarize the plan we will go over in this section to achieve custom TBA:

1. Understand and call the login endpoint of the API to fetch the access token.

2. Design and implement a page with the login form.

3. Build a JWT authentication state provider and set up the authentication infrastructure.

4. Integrate the authentication provider with the form.

5. Enable authentication for the FetchData page to test the flow.

So, let's get started.

Calling the login endpoint

As we learned in *Chapter 8*, *Consuming Web APIs from Blazor WebAssembly*, the first thing we should do before calling an endpoint is understand it. Let's run the API project and check the Swagger page to see what there is.

You can see in the **Authentication** section that there is an endpoint called /Authentication/ login with the specs, as you can see in the following screenshot from Swagger:

Figure 9.2 – Login endpoint in the BooksStore API

So, it's a simple POST request that takes an object with a username and password and then retrieves an API response object wrapping a login response object that contains the token property, which is what we want. If there is a failure, as mentioned in the *Sending a GET request* section in *Chapter 8*, all the API requests that fail will retrieve an ApiErrorResponse object that we have already created.

We can test it either with Postman or directly from the Swagger page with one of the two users I mentioned earlier. Click on **Try it out** in the top right of the request details, and then fill the request body with the following JSON:

```
{
    "username": "admin@masteringblazor.com",
    "password": "Test.123"
}
```

Click **Execute** and then, in the **Responses** section, you should see the JSON response with the token inside, as shown here:

Figure 9.3 – Login endpoint response with the access token

You can copy the token value from the response and paste it into https://jwt.ms. You can see the claims in the decoded part, as we saw in *Figure 9.1*.

Now, we have a clear idea of how this endpoint works, what it expects, and what it returns to us. Let's translate this into code that sends the request from our Blazor WebAssembly project:

1. We will start by creating the class of the object to send in the body of the request. In the `Models` folder, create a new class called `LoginRequest.cs` with two properties, as shown in the following snippet:

    ```
    using System.ComponentModel.DataAnnotations;
    using System.Text.Json.Serialization;
    public class LoginRequest
    {
        [Required]
        [JsonPropertyName("username")]
        public string Username { get; set; } =
          string.Empty;

        [Required]
        [StringLength(16, MinimumLength = 5)]
        [JsonPropertyName("password")]
        public string Password { get; set; } =
          string.Empty;
    }
    ```

    Because the same object will be used in the form we will create later, we added a data annotation for validation, such as specifying that the field is mandatory and the maximum and minimum length of the password.

2. Create another class in the `Models` folder called `LoginResponse`:

    ```
    using System.Text.Json.Serialization;
    public class LoginResponse
    {
        [JsonPropertyName("token")]
        public string? AccessToken { get; set; }

        [JsonPropertyName("refreshToken")]
        public string? RefreshToken { get; set; }
    }
    ```

3. Within the `Services` folder, create an interface called `IAuthenticationService` with one method inside:

    ```
    namespace BooksStore.Services;
    public interface IAuthenticationService
    {
        Task<LoginResponse> LoginUserAsync(LoginRequest
    ```

```
                    requestModel);
    }
```

As we said in the *Separating API calls from components* section in *Chapter 8*, we will keep the API calls wrapped inside interfaces and away from the component to keep our components clean and testable, as we will see in *Chapter 13, Testing Blazor WebAssembly Apps*.

4. Create the implementation of `IAuthenticationService`. In the `Services` folder again, create a new class called `AuthenticationService`. The class will implement the `IAuthenticationService` interface and contains the implementation of `LoginUserAsync`, which sends the requests to the login endpoint in the API, as shown here:

```
using BooksStore.Models;
using System.Net.Http.Json;

namespace BooksStore.Services;
public class AuthenticationService :
  IAuthenticationService
{

    // Inject the HttpClient into the constructor
    private readonly HttpClient _httpClient;

    public AuthenticationService(HttpClient
                                  httpClient)
    {
        _httpClient = httpClient;
    }

    public async Task<LoginResponse>
      LoginUserAsync(LoginRequest requestModel)
    {
        var response = await _httpClient
          .PostAsJsonAsync("authentication/login",
                           requestModel);
        if (response.IsSuccessStatusCode)
        {
            return await response.Content
              .ReadFromJsonAsync<LoginResponse>();
        }
        else
        {
            var error = await response.Content
              .ReadFromJsonAsync<ApiErrorResponse>();
            Console.WriteLine(error);
```

```
            throw new Exception(error.Message);
            // TODO: Handle the error in a proper way
        }
    }
}
```

So, the code inside the method is a typical HTTP request to the API. In the *Managing API errors* section in *Chapter 10, Handling Errors in Blazor WebAssembly*, we will learn how to efficiently handle those errors.

5. In `Program.cs`, register `IAuthenticationService` in the dependency injection container so we can inject it into the login component later:

    ```
    ...
    builder.Services.AddScoped<IAuthenticationService,
      AuthenticationService>();
    await builder.Build().RunAsync();
    ```

That's all for this section. Now it's time to build the login form that will call the `LoginUserAsync` method we just created.

Designing the login form page

To let users (normal customers or admins) log in, we need to provide them with a login page with a form to submit their username and password.

So, the goal of this step is to get a page that looks like the following screenshot:

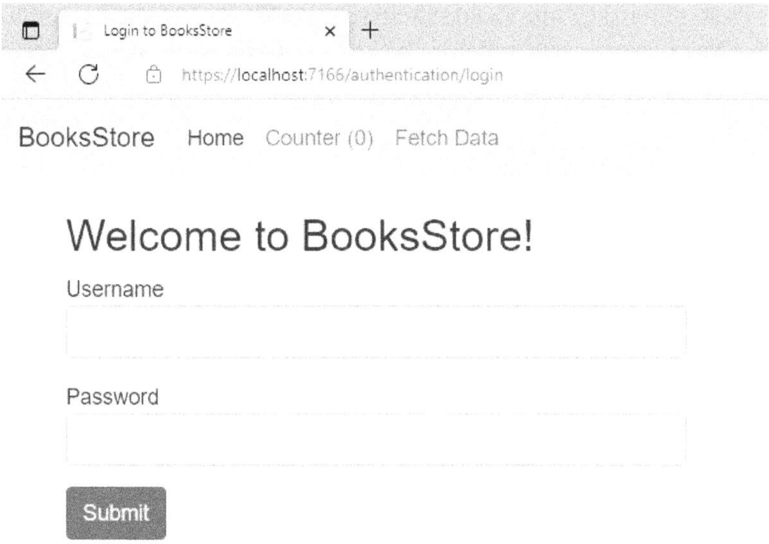

Figure 9.4 – User login page

We will start by designing the UI and then we will implement the logic for submitting the form.

Let's switch to Visual Studio and get that component done:

1. In the Pages folder, we will create another folder and call it Authentication. This is the folder where all the pages related to authentication will reside, such as for login, registration, and password reset.

2. Inside the Authentication folder, create a new Razor component and call it Login. This will represent the login page, so it needs a router and a page title, and we will apply the UserLayout for it, as shown here:

    ```
    @page "/authentication/login"
    @layout UserLayout
    <PageTitle>Login to BooksStore</PageTitle>
    . . .
    ```

3. In the @code section, create a variable of the LoginRequest type and call it _model. This will represent the object we will bind to the form. We started with this object before creating the markup because, as we learned in the *Understanding forms in Blazor* section in *Chapter 5, Capturing User Input with Forms and Validation*, the EditForm component requires its Model property to be populated to function or even be rendered in the UI:

    ```
    . . .
    @code {
        private LoginRequest _model = new();
    }
    ```

4. Now, let's create the markup with an EditForm component, DataAnnotationsValidator, two InputText components, and two ValidationMessage components to represent the login form with its validation capability:

    ```
    . . .
    <h2>Welcome to BooksStore!</h2>
    <div class="row">
        <div class="col-sm-12 col-md-4">
            <EditForm Model="_model">
                <DataAnnotationsValidator />
                <div class="mb-3">
                    <label>Username</label>
                    <InputText
                      @bind-Value="@_model.Username"
                      class="form-control" />
                    <ValidationMessage For="@(() =>
                      _model.Username)" />
    ```

```
            </div>

            <div class="mb-3">
                <label>Password</label>
                <InputText type="password"
                  @bind-Value="@_model.Password"
                  class="form-control" />
                <ValidationMessage For="@(() =>
                  _model.Password)" />
            </div>

            <button type="submit"
              class="btn btn-primary">Login</button>
        </EditForm>
    </div>
</div>
...
```

We have used some Bootstrap CSS classes for the overall design, but what matters here is how we bind the values of the two inputs to the Username and Password properties of the LoginRequest model.

Now, you can run the project; navigate to /authentication/login, and you should see a form identical to the one shown in *Figure 9.4*. The form also supports validation when you click **Login**, as we designed it with the attributes of the LoginRequest class.

After getting the design ready, the second phase is to write the logic of submitting the form. We will call the LoginUserAsync method we created, check whether the token is retrieved, and then store it in the local storage of the browser:

1. Inject IauthenticationService and ILocalStorageService:

    ```
    ...
    @using Blazored.LocalStorage
    @inject IAuthenticationService AuthService
    @inject ILocalStorageService LocalStorage
    ...
    ```

2. In the @code section, create a new method called SubmitLoginFormAsync and add the following logic to it:

    ```
    ...
    private async Task SubmitLoginFormAsync()
    {
        try
        {
    ```

```
                    // Based on the logic of the method, if the
                    // request to the API fails, it will throw an
                    // exception, otherwise it will return the
                    // token.
                    var loginResult = await
                      AuthService.LoginUserAsync(_model);
                    // Store the token in the local storage
                    await LocalStorage.SetItemAsync(
                      "access_token", loginResult.AccessToken);
                    // TODO: Trigger the Blazor app to refresh the
                    // authentication state, To be Handled in the
                    // next step
                    Navigation.NavigateTo("/");
                }
            catch (Exception ex)
            {
                    // TODO: Log the error in Chapter 11
                    Console.WriteLine(ex.Message);
            }
        }
    . . .
```

The preceding logic is straightforward: call LoginUserAsync and pass the _model used in the form as a parameter. If the token is retrieved successfully, we store it within the table storage and then navigate to the home page. If there's a failure, for the time being, we just log it into the console, but in the next chapter, we will learn how to properly deal with failures.

3. Assign the SubmitFormAsync method to the EventCallback OnValidSubmit parameter of the EditForm component in the markup section:

```
    . . .
        <EditForm Model="_model"
          OnValidSubmit="SubmitLoginFormAsync">
    . . .
```

This is everything that's needed for this step. Run the API project and the Blazor WebAssembly project, navigate to /authentication/login, and try admin@masteringblazor.com for the username and Test.123 for the password. After clicking the **Login** button, you will be taken to the **Index** page, but if you access **Developer Tools** in your browser and check the local storage, you will find the access token stored there:

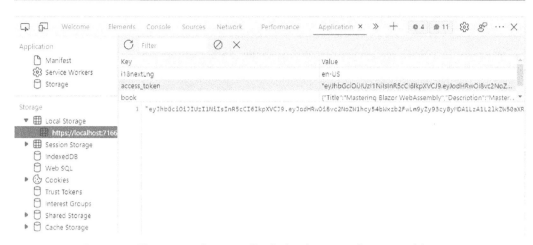

Figure 9.5 – The access token stored in the local storage after successful sign-in

So, the question now is what has changed in the system if this `access_token` value in the local storage is simply a string value? In the next step, we will decode this token, extract the claims from it, and let the Blazor app build the user identity out of this token.

Building the authentication state provider and setting up the authentication infrastructure

It's time to work on what will make the authentication happen. Everything so far has been geared toward getting the token and providing the user with the required UI to log in.

What is the general goal here? We want to implement an authentication state provider. This service will allow us to write the authentication logic we are aiming for. In our case, the authentication logic will be based on the access token value stored in the local storage. Follow these steps to process it:

1. Check whether the local storage contains the access token value.
2. Decode that access token and extract the claims from it.
3. Build a `ClaimsPrincipal` object that represents the user with its claims.
4. Notify the app about that change in the authentication state.
5. Return a new `AuthenticationState` object that contains the user.
6. If there is no token or the token is invalid for some reason, just return an empty `AuthenitcationState` object with no claims.

After building the authentication state provider, we will register it in the DI container, enable authentication in `Program.cs`, and finally, make some modifications to the `App.razor` component so our app has a full authentication mechanism.

Two packages are needed to get this work done:

- **Microsoft.AspNetCore.Components.WebAssembly.Authorization**: This is the package that has the AuthenticationStateProvider abstract class and the other authentication components.

- **System.IdentityModel.Tokens.Jwt**: This has the JwtSecurityTokenHandler class. We will use a method inside it called ReadJwtToken to parse the access token and extract the claims from it.

You can install the packages either using *NuGet* in Visual Studio or using the *dotnet CLI* via the following command:

```
dotnet add package Microsoft.AspNetCore.Components.WebAssembly.
Authorization
dot add package System.IdentityModel.Tokens.Jwt
```

Everything is ready for us now, so let's get into it:

1. In the root folder of the project, add a new class called JwtCustomAuthentication-StateProvider that inherits from AuthenticationStateProvider, add some usings to be used, and add the following code inside:

```
using Blazored.LocalStorage;
using Microsoft.AspNetCore.Components.Authorization;
using System.IdentityModel.Tokens.Jwt;
using System.Security.Claims;
namespace BooksStore;

public class JwtAuthenticationStateProvider :
  AuthenticationStateProvider
{
    public async override Task<AuthenticationState>
      GetAuthenticationStateAsync()
    {

    }
}
```

2. Inject ILocalStorageService to read the token from the local storage:

```
private readonly ILocalStorageService _storage;
public JwtAuthenticationStateProvider(
  ILocalStorageService storage)
```

```
{
    _storage = storage;
}
```

3. In the body of the GetAuthenticationStateAsync method, we will start by checking whether access_token exists in the local storage. If not, we will return an AuthenticationState object with an empty ClaimsPrincipal object inside, which will indicate that the user is not authenticated:

```
...
public async override Task<AuthenticationState>
  GetAuthenticationStateAsync()
{
    if (await
        _storage.ContainKeyAsync("access_token"))
    {
        // Process the token here
    }
    var anonymousUser = new ClaimsPrincipal(new
      ClaimsIdentity()); // Empty claims and
      // authentication scheme provided
    var anonymousAuthState = new
    AuthenticationState(anonymousUser);
    NotifyAuthenticationStateChanged(
        Task.FromResult(anonymousAuthState));
    return anonymousAuthState;
}
...
```

4. Inside the if block, we will get the access token value and parse it using the JwtSecurityTokenHandler class:

```
// Read and parse the token
var tokenAsString = await
  _storage.GetItemAsync<string>("access_token");
var tokenHandler = new JwtSecurityTokenHandler();
var token = tokenHandler.ReadJwtToken(tokenAsString);
```

5. We need to create a new ClaimsIdentity object. This object represents the user claims extracted from the token, and the name of the authentication scheme, which is jwt in our case:

```
var identity = new ClaimsIdentity(token.Claims,
                                  "jwt");
```

We can add some validation to the values of the claims to make sure the token is valid and to check for the expiration date, but to keep things simple and straightforward, I have excluded this step for now because we are going to talk about that in the last section.

6. The last thing we need to build is a `ClaimsPrincipal` object out of the identity object we just created, then a new `AuthenticationState` object, and call the `NotifyAuthenticationStateChanged` method, which is a method from the base class that will notify other components that the authentication state has changed. Finally, we need to return the `authState` object:

```
var user = new ClaimsPrincipal(identity);
var authState = new AuthenticationState(user);
NotifyAuthenticationStateChanged(Task.FromResult(
  authState));
return authState;
```

The preceding code is the engine of the authentication system that we are building.

7. In `Program.cs`, add the authentication state provider to the DI container, and also add the required authentication services by using `AddAuthorizationCore` as follows:

```
. . .
builder.Services.AddAuthorizationCore();
builder.Services.AddScoped<AuthenticationStateProvider,
JwtAuthenticationStateProvider>();
. . .
```

8. The Blazor app now knows how to authenticate our users, but components from the Blazor authentication library are needed in `App.razor` to complete the story. We need to wrap all the components in `App.razor` with the `<CascadingAuthenticationState>` component. This makes the authentication state object include the identity state and makes the claims available through a cascading parameter for any component:

```
. . .
@using Microsoft.AspNetCore.Components.Authorization
<CascadingAuthenticationState>
        <Router ...
</CascadingAuthenticationState>
```

9. Another important change in `App.razor` is replacing the `<RouteView>` component with `AuthorizeRouteView`. This component provides an authentication check for protected URLs; for example, if the user is not logged in and trying to access a protected page, `AuthorizeRouteView` allows you to customize what should happen in this case:

```
. . .
<AuthorizeRouteView RouteData="@routeData"
  DefaultLayout="@typeof(UserLayout)">
```

```
        <NotAuthorized>
            <h3>You don't have the permissions to access
                this page</h3>
            <br />
            <a href="authentication/login"
                class="btn btn-primary">Login</a>
        </NotAuthorized>
    </AuthorizeRouteView>
    . . .
```

`AuthorizeRouteView` has a `RenderFragment` parameter called `NotAuthorized`. We use it to render what will be rendered in the UI if the user is not authenticated. We also have set the default layout of all pages to `UserLayout`.

10. Many of the classes, components, and attributes in the authentication namespaces will be used in many components. So, we don't have to add the `@using` directive for those namespaces in each component. Open `_Imports.razor` in the root folder and add the following two namespaces:

    ```
    . . .
    @using Microsoft.AspNetCore.Components.Authorization;
    @using Microsoft.AspNetCore.Authorization
    . . .
    ```

11. We are almost there! The last step here is to call `GetAuthenticateStateAsync` from the authentication state provider we built within the Login page after the token is fetched and stored, which will trigger the app to evaluate the authentication state based on the stored token.

12. To do that, we just need to inject `AuthenticationStateProvider` and call the method inside the `Login.razor` page:

    ```
    . . .
    @inject AuthenticationStateProvider AuthStateProvider
    ```

    And within the `@code` section and the `SubmitFormAsync` method, we will get `GetAuthenticationStateAsync` after storing the token in the local storage and before redirecting the user to the home page:

    ```
    . . .
    await AuthStateProvider.GetAuthenticationStateAsync();
    Navigation.NavigateTo("/");
    . . .
    ```

With that, our system has a full authentication mechanism. That was a bit long, but that's how Blazor WebAssembly handles it natively. In the next section, we will see how powerful this is. It took us some time to get here, but next, we will secure some pages and some parts of the UI, as well as implement roles and policies. All of this will be easy from now on, as what we have set up so far is the core, and what's left is leveraging the available classes and components that depend on this infrastructure to complete the authentication system.

If you run the project and navigate to the login page, then enter your credentials, you will be redirected to the home page. Nothing is noticeable so far because our app doesn't have any secured pages or UI components yet.

To make sure everything is working as expected, we will try to make accessing the `FetchData` page require you to be an authenticated user. It's as simple as decorating the component with the `[Authorize]` attribute. In the `Pages` folder, open `FetchData.razor` and add the following at the top of the page:

```
@page "/fetchdata"
...
@attribute [Authorize]
...
```

When you run the app and navigate to the `FetchData` page, if you are logged in (i.e., a valid token exists in the local storage), you can see the weather forecast data normally, but if you are not logged in, you should see something like the following page:

Figure 9.6 – The unauthorized view when accessing a protected page

> **Note**
> We haven't implemented the logout feature yet. If you logged in already earlier in this chapter, you could just open the local storage in **Developer Tools** in your browser and remove the `access_token` key from there, then refresh the page, so the app will not consider you logged in.

Cool! Our authentication system is working, but is that everything? The answer is absolutely not! Blazor, with its authentication library, has too many things left to offer. In the next section, we will learn how to leverage roles and custom policies in our app, implement logout, control the render of the UI to show or hide some UI elements based on the authentication status, and much more.

Implementing authorization and advanced authentication features

We now have a new term to learn about – **authorization**. We have mentioned authentication many times, but what is authorization? It is a mechanism on top of authentication that restricts access to resources even for authenticated users.

We gave the example of an employee being allowed access to their company building because they proved that they are an employee. That's authentication. Authorization, on the other hand, means the employee cannot access all the departments inside the building. They require a special role, or permission, to do that.

In software, the same concepts apply: the user is authenticated after successfully logging in, but authorization means that the user cannot take certain actions or access certain features without a specific role or permission.

In this section, we will discover how to implement authorization using the Blazor authentication library with roles or policies. After that, we will deep-dive into some scenarios that you will encounter in almost every app you build.

Roles and policies

In the BooksStore project, we have two types of users: the system admins and the customers who want to buy books.

Admins can access the BookForm page we built in *Chapter 5* to insert a new book, and, of course, they can access anything else in the system. Customers don't have access to the BookForm page, or any feature in the system related to managing the books; they can only browse the library, buy books, and review them.

The infrastructure we have set up so far already supports roles, but the question is where are the roles of the users coming from? You know that the token has a set of claims that represents the user, and the API issues the token that contains the role claims. If you copy the access token fetched from the local storage and paste it into https://jwt.ms, you will notice the role claim there:

Figure 9.7 – Role claim in the access token for the admin user

If you try to log in with the other account, `ahmad.mozaffar@masteringblazor.com`, you will notice the role claim has a value of `Customer`. Blazor WebAssembly recognizes this as the role of the user. A token can have multiple role claims, and that's totally fine.

So, how we can take advantage of that? Well, like everything else in Blazor, it's super simple. Let's restrict the `BookForm` page so that it can be accessed by admins only. In the `Pages` folder, open the `BookForm.razor` file, and add the `[Authorize]` attribute. Populate the `Roles` property with a value of `Admin`, as follows:

```
@page "/book/form"
. . .
@attribute [Authorize(Roles = "Admin")]
. . .
```

If you have multiple roles, you can set the `Roles` property with a normal string but with a comma to separate the roles – for example, `[Authorize(Roles = "Admin,Supervisor")]`.

Run the project with the API and log in with an admin account. Navigate to `/book/form`. You will see the page normally, but if you log in again with the customer account and navigate to the same page, you will see the unauthorized view we saw in *Figure 9.6*.

Roles are great for access authorization, but that's not the only way to do it. We have a more flexible way to restrict access to the app's resources: policies. Policies can be used to get the same result, but they allow us to have more control over the requirements we want. With roles, either the user has this role or not; it's simple and straightforward. With policies, we can have more customization; for example, we can have an authorization policy that's based on the age or the country of the user, and this is very popular in many apps you use daily.

So, you can build some policies and give them the specs you are aiming for. For example, you can make a policy that is satisfied when the user has a certain claim or doesn't have another one, or when a claim has a certain value.

The token that our API retrieves has the `Country` claim, so we will build a policy based on the country. For example, our store sells only in the United Kingdom, so we need to ensure that users from outside this country won't be able to buy books or add items to their cart. Or we can restrict a new feature for a certain area for testing purposes before releasing it publicly.

To build a policy, navigate to `Program.cs` and manipulate the `AddAuthorizationCore` method we called earlier in the *Building a custom JWT authentication flow* subsection by using the overloaded version, which allows us to customize the `AuthorizationOptions` object and add policies based on that, as shown here:

```
. . .
builder.Services.AddAuthorizationCore(options =>
{
    options.AddPolicy("UK_Customer", policy =>
    {
        policy.RequireClaim(ClaimTypes.Country, "UK");
    });
});
. . .
```

The `AddPolicy` method is used to create a new policy, and it accepts a name and a delegate to configure the `AuthorizationPolicyBuilder` object. For the policy object, we use the `RequireClaim` method, which takes the name of the claim's type and the allowed value, which is UK in our example. `ClaimTypes.Country` is one of many string constants that represent predefined common claim types, and we use the `Country` claim to check for its value.

It's time to use our policy. So far, we have only restricted access to a full page using the `[Authorize]` attribute, and we did the same with roles too. With policies, it's the same thing:

```
[Authorize(Policy = "UK_Customer")]
```

For our example, we are not going to use this policy to restrict access to full pages. We will use it to prevent customers who do not fulfill the policy from adding items to their cart, which is only a specific part of the full UI, and that's what we will learn how to achieve next.

The AuthorizeView component

AuthorizeView is another powerful built-in component in the authentication library. It allows us to show or hide part of the UI based on the user's authentication state. For example, we can show the login button if the user is not authenticated but show a greeting message and a logout button if the user is authenticated. Another example we will go through is the **Add to Cart** button. We will hide this button and show a message indicating that only UK customers can buy books if the customer is not from within the UK. The **Add Book** form page now doesn't have a link in the nav bar, and we have restricted its access to only admins. We will be using `AuthorizeView` to show the link in the nav bar if the logged user is an admin.

The component takes two `RenderFragment` parameters: one called `Authorized`, which we can use to render content for authorized users, and one called `NotAuthorized`, with which we can render content for unauthorized users. Also, the authorized `RenderFragment` gives us access to the identity and the user object via the `context` parameter. If you add `AuthorizeView` with UI content directly without wrapping it with `<Authorized>` or `<NotAuthorized>`, it will render the content in `<Authorized>` by default.

With the scenarios we have covered, we can deal with almost any situation you may face in your app-building process. We can control content based on whether the user is logged in or not, based on a custom policy, or based on a role.

Let's see that in action. We will begin with the **Login** button on the right of the page. We will show the button if the user is not logged in, but we will show a message and a **Logout** button if the user is logged in:

1. In the `Shared` folder, create a new Razor component and call it `LoginDisplay`. We will have this logic in a separate component to not make the `Navbar` huge and complicated.

2. Add the following code to the component:

```
<AuthorizeView>
    <Authorized>
        @* Render the content if the user is logged in
        *@
        <div class="d-flex align-items-center">
            <p class="my-0">Hello @context.User
              .FindFirst(ClaimTypes.GivenName)?.Value
            </p>
            <button class="btn btn-danger mx-2">
              Logout</button>
        </div>
    </Authorized>
    <NotAuthorized>
        @* Render the content if the user is not
```

```
                    logged in *@
                <a class="btn btn-outline-primary"
                    href="/authentication/login">Login</a>
            </NotAuthorized>
        </AuthorizeView>
```

In the preceding code, `AuthorizeView` is very straightforward. In the `<Authorized>` parameter tag, we add the greeting message and access the user claims using `@context` with a **Logout** button. We will write its logic later.

3. In `NavBar.razor`, also in the `Shared` folder, in the last div, we have a **Login** button. We will replace it with the `LoginDisplay` component:

    ```
    ...
    <div class="d-flex">
        <LoginDisplay />
    </div>
    ...
    ```

Let's run the app and see the result in both cases. If you are not logged in, you will see the normal **Login** button, but this time, it will direct you to the **Login** page. Otherwise, you should see the result shown in the following screenshot:

Figure 9.8 – Content if the user is logged in

The next scenario we want to cover is using `AuthorizeView` to render content based on a specific role. We will add a link to the `BookForm` page in `NavBar.razor` as a new `<li>` HTML tag below the existing links, but it will only appear to admin users. We have two ways to achieve this. The first is to use the `IsInRole` method for the `User` object within the `<Authorized>` parameter tag, as demonstrated here:

```
...
<AuthorizeView>
    <Authorized>
        @if (context.User.IsInRole("Admin"))
        {
            <li class="nav-item">
                <NavLink class="nav-link"
                    href="/Book/Form">Add Book</NavLink>
            </li>
        }
```

```
            </Authorized>
    </AuthorizeView>
    ...
```

Alternatively, we can achieve the same result directly using the `AuthorizeView` component:

```
<AuthorizeView Roles="Admin">
    <li class="nav-item">
        <NavLink class="nav-link" href="/Book/Form">
            Add Book</NavLink>
    </li>
</AuthorizeView>
```

So, which one should you use? Both achieve the same goal. The second one is more readable and has less code, but the first one gives you more manipulation power and flexibility. Also, if you want to render special content for unauthorized users, you need to add both tags, `<Authorized>` and `<NotAuthorized>`.

You will see the **Add Book** link at the top when you run the app if you are logged in using an admin account. Otherwise, the link will not be there.

The last thing to talk about for `AuthorizeView` is rendering content based on a policy. We have created the `UK_Customer` policy, so let's apply it to the `Add to Cart` button within the `BookCard` component in the `Shared` folder:

```
...
<AuthorizeView Policy="UK_Customer">
    <Authorized>
        <button class="main-button"
            @onclick="AddToCart">Add to Cart</button>
    </Authorized>
    <NotAuthorized>
        <p class="text-center">Books only available for
            sale in UK for the time being</p>
    </NotAuthorized>
</AuthorizeView>
...
```

Easy and powerful, right? Let's test it and make sure it's working as expected by logging in and using the `ahmad.mozaffar@masteringblazor.com` account. This customer is located in the *UK*, so we should see the button, but because we're using the admin account, we should see something like this:

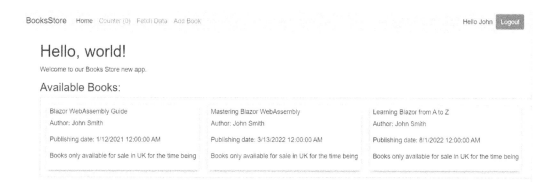

Figure 9.9 – Book card without an Add to Cart button

That's all for `AuthorizeView`. The scenarios we have covered allow you to customize your user experience up to the enterprise level. By setting up your roles and the policies correctly, you will be able to satisfy any requirement from small use cases up to enterprise-level scenarios.

CascadingAuthenticationState

We have seen how to control the UI for either a full page or part of it based on the authentication state, but what about the logic of the app? What if we want to retrieve the previously added user's shopping cart items from the API? This action should be taken if the user is logged in only; otherwise, it's not needed.

Earlier, in the *Building a custom JWT authentication flow* section, we modified `App.razor` and wrapped all the components with a `<CascadingAuthenticationState>` component. All this component does is provide an `AuthenticationState` object to the child components via a cascading parameter.

If we want to run specific logic only if the user is authenticated in the `Index.razor` page after the component is initialized, all we need to do is follow these steps:

1. Open `Index.razor` in the `Pages/UserPages` folder.
2. Add a cascading parameter of the `Task<AuthenticationState>` type to the @code section:

    ```
    ...
    [CascadingParameter]
    public Task<AuthenticationState>? AuthenticationState
        { get; set; }
    ...
    ```

3. In the `OnInitializedAsync` life cycle method, you can get the state and the user object and write your logic just like this:

```
...
var authState = await AuthenticationState;
if (authState.User.Identity.IsAuthenticated)
{
    // Execute logic if the user is authenticated
}
...
```

The same concepts are applicable when it comes to roles or reading the claims and so on, as we learned earlier in the chapter.

Sign-out functionality

Of course, you don't want the app to allow users to log in and not allow them to log out again. Our authentication uses the token in the local storage to determine the authentication process. If we remove that token from the storage and tell `AuthenticationStateProvider` to evaluate the authentication state again, there will be an empty authentication state with no claims or authentication scheme, which means no user is logged in.

We will apply the logout logic to the `LoginDisplay.razor` component we created in the *The AuthorizeView component* subsection part earlier in this section:

1. Inject the `ILocalStorageService` and `AuthenticationStateProvider` instances:

```
@using Blazored.LocalStorage;
@inject ILocalStorageService LocalStorage
@inject AuthenticationStateProvider
  AuthenticationStateProvider
...
```

2. In the `@code` section, add the `LogoutAsync` method. It will remove the token and let `AuthenticationStateProvider` fetch the new state again. It will return an empty authentication state and notify the whole app about this change:

```
...
private async Task LogoutAsync()
{
    await
      LocalStorage.RemoveItemAsync("access_token");
    await AuthenticationStateProvider
```

```
        .GetAuthenticationStateAsync();
    }
    ...
```

3. Assign the `LogoutAsync` method to the `@onclick` event of the **Logout** button in the `AuthorizedView` component:

    ```
    <button class="btn btn-danger mx-2" @
    onclick="LogoutAsync">Logout</button>
    ```

That's it! Run the app and give it a test. Click the **Logout** button and, if you are logged in with the admin account, you will see that the **Add Book** link in the navbar will disappear. The greeting message with the **Logout** button will disappear too, and the **Login** button will appear again.

Auto-redirect to login

When the user is not logged in and opens the `FetchData` page, only content that is not authorized will be rendered. We defined the content in `App.razor` within the `AuthorizeRouteView` component.

This is one approach. The other way is to redirect the user to the login page automatically. To have better control over what the user sees while accessing an authorized page, whether we're controlling the UI or the logic, it's recommended to create a separate component that renders and does the logic we are aiming for and render this component in `AuthorizeRouteView`.

Let's implement auto-redirect to the login page step by step:

1. Create a component called `RedirectToLogin` in the `Shared` folder.

2. Inject `NavigationManager` and, in the `@code` section, override the `OnInitialized` method and navigate the user to the login page:

    ```
    @inject NavigationManager Navigation
    @code {
        protected override void OnInitialized()
        {
          Navigation.NavigateTo("/authentication/login");
        }
    }
    ```

    The component is very straightforward. It doesn't need to render anything. Whenever it's loaded, it will navigate the user to the login page.

3. Add the `RedirectToLogin` component inside the `<NotAuthorized>` tag in `AuthorizeRouteView` within `App.razor` instead of the text and the login button:

    ```
    ...
    <AuthorizeRouteView RouteData="@routeData"
    ```

```
    DefaultLayout="@typeof(UserLayout)">
      <NotAuthorized>
          <RedirectToLogin />
      </NotAuthorized>
  </AuthorizeRouteView>
  . . .
```

By following this method, you have more flexibility over what should happen if a user accesses a protected page. This intermediary component can do more than `RedirectToLogin`, but we have demonstrated that as it's a common flow.

With that, we have come to the end of exploring Blazor's authentication capabilities and how to put them into practice. However, some or all the API endpoints you are dealing with can be authorized, so you need to send the access token in the HTTP request while accessing those endpoints. This is what we will learn about in the upcoming section.

Accessing authorized API endpoints

In the API of `BooksStore`, or any APIs you are working with, either some or all the endpoints will be secured and will require an authorized user with the right roles and permissions. In this book, we are looking at web APIs from the client's perspective, so we are not going to explain how the API authentication system works internally. The API documentation should tell you how the authentication works and what you should do to access the protected endpoints; if you are the developer of the API, you should know how it works directly.

In this chapter, we used JWT authentication, which is the most common form of authentication nowadays, even if you are dealing with identity providers such as Microsoft, Google, and Facebook, rather than the custom flow that we built; they all work the same. With this authentication scheme to access a protected endpoint in the API, all you should do is send the access token that the API gave to you in the login process within each HTTP request you are sending for that API. That sounds easy, and it is easy actually, and we will see how to do that practically in Blazor WebAssembly next.

In the flow I just mentioned, whenever we are using `HttpClient` to send an API request, we can grab the token from the local storage, attach it to the header of the HTTP request, and we're done. But that will make us write the same code every time we add a new HTTP request to the API. What we will do instead is rely on the *HTTP Message Handlers*, which we discovered in the *Exploring IHttpClientFactory and delegating handlers* section in *Chapter 8*.

An HTTP Message Handler specifically for authorization, every time we send a request to the API, automatically gets the token from the storage and puts it in the header of the request. By following this approach, we will write this logic only once, and it will apply to every HTTP request being sent, as well as the ones you will add later.

For the `BooksStore` API, so far, we only have the `POST` request to add a new book that is authorized. If an HTTP request is sent without a token, with an invalid token, or with a token but not for an admin user, the API will return a `401` status code, which means the request is not authorized.

If you want to test that flow in Postman, you can get a token using the login endpoint, then copy that token and set it in the authorization header value using the **Authorization** tab, as illustrated in the following screenshot:

Figure 9.10 – Setting the JWT in the header of the request

Here, Postman is adding a new header value to the HTTP request with the `Authorization` key, and its value is the word `Bearer` concatenated with the user token.

We will build the message handler and, inside it, we will simulate what Postman is doing for the `HttpRequestMessage` object that is being sent to the API:

1. In the root folder, create a new class called `AuthorizationMessageHandler.cs`.

2. Let the class inherit from `DelegatingHandler` and inject `ILocalStorageService` so we can fetch the token:

```
using Blazored.LocalStorage;
namespace BooksStore;
public class AuthorizationMessageHandler :
  DelegatingHandler
{
    private readonly ILocalStorageService
      _localStorage;
    public AuthorizationMessagingHandler(
```

```
                ILocalStorageService localStorage)
        {
            _localStorage = localStorage;
        }
    }
```

3. Override the `SendAsync` method. It checks whether the token is there because the token will not be in the local storage if the user is not logged in. After that, attach it to the header of the request, as the following code snippet shows:

```
    ...
    protected override async Task<HttpResponseMessage>
      SendAsync(HttpRequestMessage request,
              CancellationToken cancellationToken)
    {
        if (await
          _localStorage.ContainKeyAsync("access_token"))
        {
            var token = await _localStorage
              .GetItemAsync<string>("access_token");
            request.Headers.Authorization = new
              System.Net.Http.Headers
                .AuthenticationHeaderValue("Bearer",
                                            token);
        }
        return await base.SendAsync(request,
          cancellationToken);
    }
    ...
```

4. Register `AuthorizationMessageHandler` in the DI container and attach it to the `HttpClient` instance we have in `Program.cs` using the `AddHttpMessagingHandler` method:

```
    ...
    builder.Services.AddScoped<AuthorizationMessageHandler>();
    builder.Services.AddHttpClient("BooksStore.ServerAPI", client =>
    client.BaseAddress = new Uri(builder.Configuration["ApiUrl"]))
        .AddHttpMessageHandler<AuthorizationMessageHandler>();
    ...
```

Another task was accomplished, and as usual, we need to make sure that it is working properly. To validate the token being sent in every request, we need to monitor the HTTP calls being sent from the app using the browser's **Developer Tools**. Run the app, open **Developer Tools** in the browser, and navigate to the **Network** tab. Then, log in to the app while the developer tools are open, and finally, navigate to the **Index** page. Because the **Index** page is sending a request to the API to fetch the books, we should see the request in the **Network** tab, and we should see the token in the header of the request too.

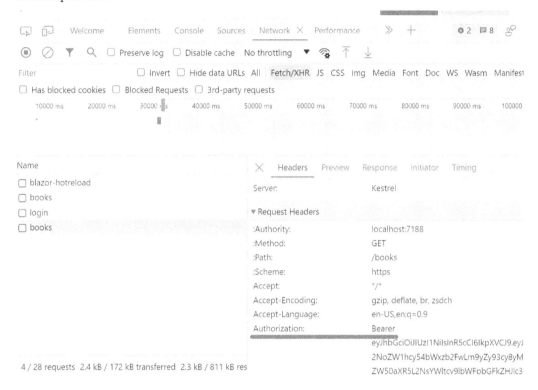

Figure 9.11 – The Bearer token sent in the authorization header of the HTTP request

When the user is not signed in, you will notice that the same request will be sent but with the `Authorization` header value.

That's how you can call an authenticated API endpoint. If this token is valid and satisfies the authorization requirements of the API endpoint, you will get the data or execute the action successfully. Otherwise, the API will send you a response with a status code of `401`.

We have done a lot in this chapter up to this point, but Blazor still has a lot to offer. In the next section, we are going to talk about what Blazor still provides out of the box for other scenarios, such as authentication with Microsoft Azure AD or AD B2C, which is part of the **Microsoft Identity Platform**.

Integrating with the Microsoft Identity Platform

We have taken advantage of many features the Blazor authentication library provides. We have implemented all those features step by step and from scratch. Microsoft has some of the best identity solutions on the market, though: **Azure AD** and **AD B2C**. What's great about it is that Microsoft also provides a library for all possible clients (JavaScript frameworks, Xamarin, .NET MAUI, native Windows frameworks, and, of course, Blazor). The client library Microsoft provides is called the **Microsoft Authentication Library** (**MSAL**). When it comes to Blazor WebAssembly, MSAL is built on top of the Blazor authentication library and already provides much of what we have done out of the box. What's even better is that .NET provides a couple of project templates that allow you to create the Blazor app with authentication and MSAL set up for you, so you just build components and the business logic because everything else is ready.

In this section, we won't write code or explain AD in depth because that would require a whole other chapter or even a full book, but we will introduce it in a nutshell so that you'll know whether it suits your needs or not. Luckily, the Microsoft docs have some very concise information on the subject, but I will provide some explanation of the process overall and when it's needed.

What is Azure AD?

Briefly, Azure AD is a cloud-based identity provider developed and maintained by Microsoft. Recently, Microsoft rebranded its identity platform as **Microsoft Entra**, so Azure AD is part of the **Entra** platform. Azure AD provides full access management features for organizations. Many companies use Microsoft 365 products, and you can access those products using what's called a Microsoft work or school account, which is the same as an Azure AD user account. Personal Microsoft accounts, such as the one you use for Outlook or Xbox, are available on Azure AD too. When using Azure AD for your solution, your app cannot register users. It will just allow them to log in using their existing Microsoft personal or work and school account, depending on your scenario.

What is Azure AD B2C?

Azure AD B2C is another identity and access management service from Microsoft, but this one differs a lot from the standard Azure AD offering. Azure AD B2C is ideal for software solutions that have users whom you want to have full control over and whom you want to allow to create accounts, log in, integrate with many social providers, edit profiles, and so much more. Azure AD B2C is capable of handling millions of users at scale and the billions of daily authentications made by these users. It enables unlimited customization for the user fields, the design and branding of the login and register pages, token claims, and more.

The difference between Azure AD and Azure AD B2C is that Azure AD users and accounts are Microsoft accounts, but you can integrate your app with them, while with Azure AD B2C, Microsoft provides you with all the identity infrastructure and tools while you own the users, customize the login and register flows, and customize the branding. As the name indicates, B2C means there is a business, such as the `BooksStore` library in our example, and there are users for it, while in Azure AD, there are Microsoft accounts and you integrate with them.

Another common service that many developers know that is similar to B2C is **Auth0**. You can learn more about it at `https://auth0.com`.

When should I use Azure AD, Azure AD B2C, or a custom flow?

The answer varies depending on the scenario. If you are the full owner of the solution or are the decision-maker, then the following tips will help, but if you are only a frontend developer on the system, the backend team and the architects will be better placed to decide what to do.

In some scenarios, such as legacy solutions, you don't have a choice. The app is there, and a database full of millions of users has been up and running for years. That means you need to build a custom authentication flow unless the team wants to migrate to a service such as Azure AD B2C.

If your app is built for organizations that use Microsoft 365 for a single organization or even for what is known as a multi-tenant app, integrating with Azure AD is the best option because, with MSAL, you can get an API and Blazor WebAssembly with authentication set up without writing any code. Another situation in which you should use Azure AD is when your app targets Microsoft service APIs such as Microsoft Graph API, for instance. In this situation, you must use Azure AD.

The most common scenario is that you are building **Software as a Service (SaaS)**: either software to be built by you for public users, or software for a single organization that has users. Azure AD B2C is the best option here. For example, let's say you are building an e-commerce platform, a multiplayer game, or a calendar app. I highly recommend going with B2C not only for its high security or scalability, which are huge benefits, but also because of MSAL and the Blazor WebAssembly project templates, which create the project for you with all the authentication needed, even with an API.

> **Note**
>
> To build a new Blazor WebAssembly app secured with Azure AD B2C, you can refer to the steps at `https://learn.microsoft.com/en-us/aspnet/core/blazor/security/webassembly/standalone-with-azure-active-directory-b2c?view=aspnetcore-7.0`.
>
> For apps with the normal Azure AD, visit `https://learn.microsoft.com/en-us/aspnet/core/blazor/security/webassembly/standalone-with-azure-active-directory?view=aspnetcore-7.0`.

We discussed a lot while building our custom flow, but that was intentional. Now, even if you have created a Blazor WebAssembly project with MSAL, everything is ready. With the knowledge gleaned from this chapter, you should know exactly what's happening under the hood and be able to easily manage special authentication scenarios, and that was the main goal of this chapter.

Summary

In this chapter, we saw how to get the best out of Blazor WebAssembly when it comes to authentication, from making the API call and getting the token to setting up the authentication infrastructure, such as the authentication state provider and the authentication components. After that, we covered as many scenarios as possible that you may face in your real-world applications, such as role-based authorization and custom policies, and how to integrate them with built-in authentication components to secure and control the UI and the logic of the app based on the user identity.

We ended the chapter with a brief introduction to the Microsoft Identity Platform with Azure AD and Azure AD B2C, in addition to MSAL and what it provides out of the box when it comes to Blazor WebAssembly.

After completing this chapter, you should be able to do the following:

- Understand the client-side authentication flow for single-page applications
- Build a powerful authentication mechanism for your Blazor WebAssembly app
- Set up role-based authorization or a custom authorization flow using authorization policies
- Control the UI components, pages, and logic based on user identity
- Be aware of the Microsoft identity systems and when to choose them

When we have users in our system, and we can communicate with our API, it's a good time to increase the reliability of our project by having a strong error-handling mechanism, which we'll cover in the next chapter. The user won't see our app crash, and we will learn how to handle everything gracefully.

Further reading

- Securing the Blazor WebAssembly project: `https://learn.microsoft.com/en-us/aspnet/core/blazor/security/webassembly/?view=aspnetcore-7.0`
- The Blazor Authentication Library: `https://learn.microsoft.com/en-us/aspnet/core/blazor/security/webassembly/standalone-with-authentication-library?view=aspnetcore-7.0&tabs=visual-studio`
- Blazor WebAssembly authentication with Microsoft accounts: `https://learn.microsoft.com/en-us/aspnet/core/blazor/security/webassembly/standalone-with-microsoft-accounts?view=aspnetcore-7.0`
- Blazor WebAssembly authentication with Azure AD: `https://learn.microsoft.com/en-us/aspnet/core/blazor/security/webassembly/standalone-with-azure-active-directory?view=aspnetcore-7.0`
- Blazor WebAssembly authentication with Azure AD B2C: `https://learn.microsoft.com/en-us/aspnet/core/blazor/security/webassembly/standalone-with-azure-active-directory-b2c?view=aspnetcore-7.0`

10

Handling Errors in Blazor WebAssembly

You are working hard on the client side, you have built an amazing app that does everything perfectly, the UI is great, your app is fetching and sending data from and to the API correctly, and everyone is happy. But what if the API, for some reason, is taking too long to process requests, so your app loads for around 30 seconds, then a yellow bar appears at the bottom, indicating that an unhandled exception occurred? That's an experience that neither the user nor the developer would like to have.

Errors and exceptions can occur for many reasons and in many parts of an app. The goal of this chapter is to learn the techniques to handle them properly and give users a better experience. We will start by explaining why error handling matters and then how to manage the errors that could arise from communication with the API. Then, we will build a global error handler, where all errors will be handled and processed properly. Finally, we will discuss the ErrorBoundary component introduced in .NET 6.0 and see how it can be used for error handling.

The main topics that this chapter will address are as follows:

- Understanding error handling
- Managing API errors
- Implementing global error handlers
- Utilizing the ErrorBoundary component

Technical requirements

The Blazor WebAssembly code and the API project used throughout this chapter are available in the book's GitHub repository in the Chapter 11 folder:

https://github.com/PacktPublishing/Mastering-Blazor-WebAssembly/tree/main/Chapter_10/

Make sure to have the API solution up and running beside the Blazor one.

Understanding error handling

Failures can show up in many parts of the app, either due to some lack of handling for the business logic or due to external circumstances such as the API being unreachable. The process of reacting to those errors gracefully without letting them crash the app is what's called error handling.

Errors don't always imply bad code or incorrect logic; these are just some of the causes of errors. Sometimes, the business logic of the app forces an error to be shown if the user is violating the rules. Other errors can be caused by external circumstances, such as the API being down. Writing good code can prevent programming errors. For example, if you are accessing a null object, that will throw a `NullReferenceException`, but this exception is avoidable with better code management.

Let's take an example of a feature that imports data into the system. The user can upload a file, and the logic of the feature forces this file to be either an Excel sheet or a CSV file. From a programming perspective, there is nothing wrong with loading any file, but from a logic point of view, only Excel or CSV files are allowed. So, the app shouldn't allow the user to pick an image, for instance; at the same time, the app shouldn't crash. If the user managed to load an improper file, the app simply shouldn't proceed with processing it and should tell the user, with a clear message, that only Excel and CSV files are allowed.

Handling errors correctly is always associated with the robustness of the system. Even if you build an app that's full of amazing features but the logic is prone to failure, your users won't be happy, and neither will you. Errors and exceptions are a source of stress for the developers in the production environment. Following the standards and maintaining proper error processing can boost your confidence in what you have developed, make it easier for you to detect errors, and keep your users happy.

Now the question is: what measures should we follow to keep our app healthy and efficient to avoid or handle errors?

Well, there are no magic solutions or components to prevent errors in the app, but there are some practices that you can follow to help reduce errors or make your app react better to them when they occur:

- Following best coding practices
- Giving exception scenarios in your logic a lot of attention
- Taking compiler errors seriously
- Going over open source code can dramatically increase your code efficiency
- Writing unit tests, as we will learn in *Chapter 13*, can detect errors in the development stage
- Having proper validation for your inputs
- Having a deep understanding of the features rules and business logic can help you identify scenarios early that could lead to potential errors later

As I have mentioned, following the best practices will reduce the number of errors but won't prevent them, so we need a good mechanism to react to those errors, and that's what we will be doing in the next three sections.

Back in our BooksStore project, right now, the app does some cool stuff and has a good authentication system, as well as an advanced form for adding books, but it's so easy to break it. We can easily see how: just run the app without running the API project. That will simulate the API's unavailability. You will notice immediately the light yellow bar at the bottom of the screen that indicates that an exception has occurred.

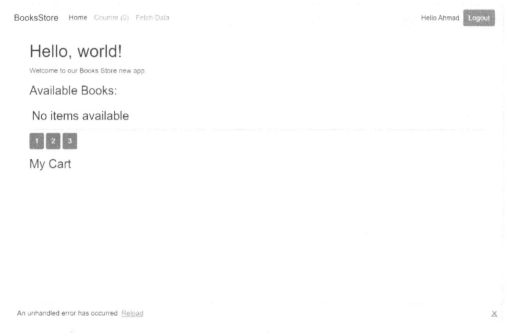

Figure 11.1 – Blazor error bar for unhandled exceptions

We need our users to have the best experience, and even when such an incident occurs, the app should indicate what went wrong in a better way. So, the plan we will follow to achieve that is as follows:

- Understand the API endpoints for failure responses
- Handle API scenarios such as network failures, internal server errors when it returns the 500 status code, and unauthorized requests scenarios
- Build a global error handler that centralizes logging and exception handling
- Use the ErrorBoundary component to show a specific error message when an exception happens while rendering a component

In the next section, we will start with the points related to the API; let's get into it.

Managing API errors

APIs are the main source of headaches in the project. API calls happen through the network and the potential for failure always exists. The API could respond differently based on the status of the response, and other scenarios could occur, such as the server being down, unhandled exceptions occurring while processing the request, or authentication and authorization errors.

For all these reasons, it's very important to handle API requests carefully. The starting point is to understand at each point what could go wrong and what the appropriate response is in each case. If you are the developer of the API, things are easier, but if you are not, the documentation should be the main source of truth.

The `BooksStore` API has the Swagger page that we used in previous chapters to understand the input and output of some API endpoints. We noticed that the endpoints either succeed or retrieve the `400 Bad Request` status code, which means something went wrong on the client side. Errors with `400` responses don't mean programmatical errors. For example, if you pass the wrong username and password to the login endpoint, the API will return a `400 status` response with an `error` object indicating that the credentials are invalid. The app should handle these scenarios properly and reflect what happens to the user, in contrast to other errors, such as network failures, where you should just let the user know that something went wrong while connecting to the server.

For the `BooksStore` API, if there is an error, all the endpoints return the same object, an object containing a message, and an array of errors.

There is no single best approach to handling this task. Some developers prefer to return a generic C# object called `Result<TSuccess, TFailed>`, for example. The object contains both properties for the objects in the success response and the failure response and then uses the `Result` object to control how the app reacts to each status. Another solution, which is the one we will use, is throwing a custom exception that contains the error object coming from the API; then, if this exception occurs in the logic of the component, we show an appropriate message or react accordingly.

The procedure we will follow relates to responses with a `400` status code. We will throw a custom exception called `ApiResponseException`. This exception will hold the error object coming from the API but doesn't imply something went wrong programmatically; it means the API is clarifying what went wrong with the request, so we should show the message to the user as it was retrieved from the API. For other response types, such as `404 Not Found`, `403 Forbidden`, or `500 Internal server error`, we will throw an exception for the time being telling the user that something is wrong and to please try again later.

> **Note**
>
> Many APIs use error codes in the `response` object; so, instead of returning a message that says `Invalid username or password`, it returns an error code with the `INVALID_ CREDENTIALS` value. Handling error codes is great and makes it easier to build the error messages in the client by mapping each error code to a message, which could be a localized message or even an action to take for a specific error code.

Because this book is looking at the API from the client's perspective only, we will not go into the solutions for keeping your API available and robust against failures. Also, you may not be using your own API; your app may rely on third-party APIs. So, anything can happen, and our client app should react correctly in any possible scenario.

So, let's approach the solution step by step:

1. Create a new folder in the project called `Exceptions`, and inside it, create a new class called `ApiResponeException.cs`.

2. The class must inherit from the `Exception` class and should have a property of `ApiErrorResponse`, the model we created in *Chapter 9* for the failure cases, and it should pass the message to the constructor of the base class, as shown here:

    ```
    using BooksStore.Models;
    namespace BooksStore.Exceptions;
    public class ApiResponseException : Exception
    {
        public ApiErrorResponse ErrorDetails { get; set; }
        public ApiResponseException(ApiErrorResponse errorDetails) :
    base(errorDetails.Message)
        {
            ErrorDetails = errorDetails;
        }
    }
    ```

    If the API doesn't return a unique object type for error cases, you may consider using the generic `ApiResponseException<TError>`, where `TError` is the type of error for each specific call.

3. Now, we can throw this exception and handle other errors for each API request. So far in our system, we have made three API calls: one to add a new book, another one to retrieve the books, and the last one to fetch the access token. Let's implement this handling for the login request within `AuthenticationService.cs` in the `Services` folder. Inside the `LoginUserAsync` method, we only handle success cases and treat everything else the same way. We will modify it to be as follows:

    ```
    using BooksStore.Exceptions;
    ...
    ```

```
var response = await _httpClient.
PostAsJsonAsync("authentication/login", requestModel);
if (response.IsSuccessStatusCode)
{
    return await response.Content.
ReadFromJsonAsync<LoginResponse>();
}
else if (response.StatusCode == System.Net.HttpStatusCode.
BadRequest)
{
    // Handle the bad request as the API doc says
    var error = await response.Content.
ReadFromJsonAsync<ApiErrorResponse>();
    throw new ApiResponseException(error);
}
else
{
    // Throw exception for other failure responses
    throw new Exception("Opps! Something went wrong");
}
...
```

You can be very detailed and handle each possible response code in a special way, and you can use the ILogger service to log errors other than 400 Bad Request:

```
else
{
    var content = await response.Content.ReadAsStringAsync();
    _logger.LogError($"Failed to log the user in. Status code
{response.StatusCode}", content);
    // Throw exception for other failure responses
    throw new Exception("Opps! Something went wrong");
}
```

The API docs will help you understand what to expect in case of failures, or if you are already the developer, then you know the exact behavior in the failed scenarios. You can apply the same concept we used for all other API calls.

4. Now, we need to handle the exceptions that could arise from the API call in the Login component. Open Login.razor within the Pages/Authentication folder and add a string variable called _errorMessage. This variable will be rendered in the UI below the **Login** button, but only if it's not empty:

```
...
<button type="submit" class="btn btn-primary">Login</button>
    @if (!string.IsNullOrWhiteSpace(_errorMessage))
    {
        <div class="alert alert-danger my-2">
```

```
                @_errorMessage
            </div>
        }
    </EditForm>
        </div>
</div>


@code {
...
        private string _errorMessage = string.Empty;
...
```

5.   In the `SubmitLoginFormAsync` method, we already have a `try`/`catch` block wrapping the call, but we will add another `catch` to handle `ApiResponseException` if it occurs. For this exception, we need the user to see what went wrong, such as invalid credentials. So, we will populate `_errorMessage` with an exception message. Also, when the user clicks **Login**, we will clear `_errorMessage` at the beginning of the submit form logic:

```
private async Task SubmitLoginFormAsync()
{
    _errorMessage = string.Empty;
    try
    {
        ...
    }
    catch (ApiResponseException ex)
    {
        _errorMessage = ex.Message;
    }
    catch (Exception ex)
    {
        // TODO: Log the error in Chapter 11
        Console.WriteLine(ex.Message);
    }
}
```

That's it for now for the general exception handling. We will be visiting this again in the next section after building the global error handler for those cases.

To test what we have done, run the project alongside the API. The login endpoint in the API will return `400 Bad Request` if you supply an invalid username and password. So, try to log in with random values and you should see the following:

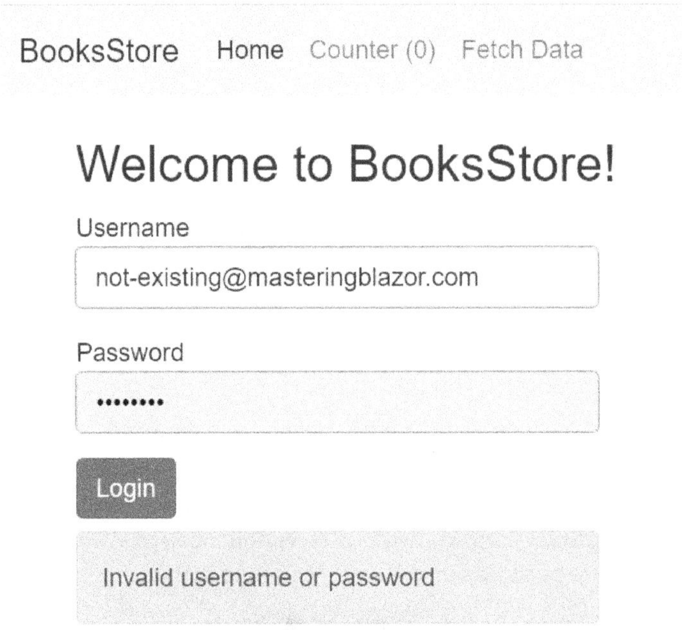

Figure 11.2 – Login with error handling

That's cool, we resolved one flow. This flow is important because it's part of the business logic and is quite common. It's very important to handle 400 errors carefully. The same approach applied here must be applied to every API endpoint you call based on its specification.

Next, we will get into the general exceptions and the rare cases. We have the function to catch general exceptions; they are not processed right now. The message is just written in the console. In the next section, we will develop a general approach to handling all such errors from a single place.

Implementing global error handlers

Having a global error handler gives you a centralized place to handle all errors the same way. This will also allow centralized logging. Errors such as API request timeouts, internal server errors, or any other unexpected scenarios will be handled here. Handling them will consist of showing a readable message to the user, indicating that something went wrong and logging the error.

We will use cascading parameters to create an error handler component for our app. This component will be placed in the App.razor file and will wrap all the other app components. It will have a method that takes an exception as a parameter and processes it accordingly. By using a cascading parameter, we can provide the error handler component as an object for all the descendant components in our app. With that, any component can have access to the handler instance and will be able to call the handle method exception.

Let's implement the flow and use it within the `Login` component:

1. In the `Shared` folder, create a new component called `ErrorHandler.razor`.

2. When an exception happens, we need to show the user a popup or message somehow. A common UI scenario is showing what's known as a *snackbar* in the top right for a couple of seconds, then it fades out. It's all up to your UI design and user experience. We will mention some UI frameworks in *Chapter 15, What's Next?*. For now, to keep things on point, we will use the default browser alert dialog by invoking the JavaScript `alert` method. To do that, we inject the `IJSRuntime` service and create a method called `ShowAlertAsync` in the code section, as shown here:

```
@inject IJSRuntime JS

@code {
    private Task ShowAlertAsync(string message)
    {
        await JS.InvokeVoidAsync("alert", message);
    }
}
```

3. The component won't have any UI markup, so it will just render the child components via a `RenderFragment` and wrap it with the `<CascadingValue>` we learned about in the *Passing data between components* section of *Chapter 2*:

```
. . .
<CascadingValue Value="this" IsFixed="true">
    @ChildContent
</CascadingValue>
@code {

    [Parameter]
    public RenderFragment ChildContent { get; set; }
    . . .
}
```

4. Inject the `ILogger` service and write the method that will process the exceptions. We will call it `HandleExceptionAsync`. The method will show a message based on the type of exception. We can use the C# 7.0 patterns feature in the `switch` statement so we can process the exception object based on its type. This way, we will be able to process the exceptions more specifically and show relevant messages:

```
@inject ILogger<ErrorHandler> Logger
. . .
public async Task HandleExceptionAsync(Exception ex)
{
```

```
        Logger.LogError(ex, ex.Message);
         switch (ex)
         {
            // TODO: Handle more specific exception
            case HttpRequestException _:
            await ShowAlertAsync("Failed to connect to the server.");
                break;
            default:
                await ShowAlertAsync("Something went wrong!");
                break;
         }
    }
    ...
```

The method is simple: it logs the error to the console window of the browser and checks the type of the exception. If it's an HttpRequestException type, we show the user an error related to the API connection. If it's the base Exception type, we just show that something went wrong. You can add more special handling for different types of exceptions in the same way as we did for HttpRequestException.

5. Before we can use ErrorHandler, we need to add it to App.razor so it wraps all the app components to provide them with its cascading value:

    ```
    ...
    <ErrorHandler>
            <CascadingAuthenticationState>
                ...
            </CascadingAuthenticationState>
    </ErrorHandler>
    ```

6. Now, we can use our error handler. In the Login page component within the Pages/ Authentication folder, add an ErrorHandler cascading parameter and call it ErrorHandler:

    ```
    ...
    @code {
        [CascadingParameter]
        public ErrorHandler ErrorHandler { get; set; }
        ...
    }
    ```

7. In SubmitLoginFormAsync within the second catch for the base Exception type, remove the Console.WriteLine call, execute the HandleExceptionAsync method from ErrorHandler, and pass the ex exception object as an argument:

    ```
    ...
    catch (Exception ex)
    ```

```
    {
        await ErrorHandler.HandleExceptionAsync(ex);
    }
    ...
```

This can be applied to every method that could fail. In the project folder cloned from the GitHub repo of the project, you will find the same technique applied to all components that use the API calls.

> **Note**
>
> The `ILogger` service will log any messages and errors to the console window of the browser. So, in a production environment, if an error occurs on a user's device, you won't know without accessing the user's browser logs. To send the logs to your server, consider using the popular logging package **Serilog** and setting it up to sink the logs to your API, as explained at `https://nblumhardt.com/2019/11/serilog-blazor/`.

To make sure `ErrorHandler` is working as expected, run the project without running the API and try to log in. After a second, an alert will appear indicating that the app is not able to connect to the server because the `HttpClient` instance throws a `HttpRequestException`, as the following screenshot shows:

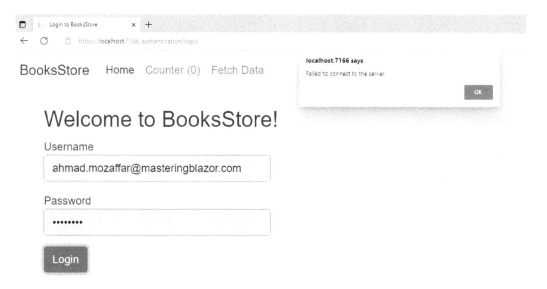

Figure 11.3 – ErrorHandler displaying an error alert

If you communicate with an online API and you don't have control over its availability, you can simulate network issues in your app using **Developer Tools** in your browser. You can set the network throttling mode to **Offline**. The following screenshot is taken from the *Edge* browser, but the same feature is available in *Chrome*, *Firefox*, and other browsers:

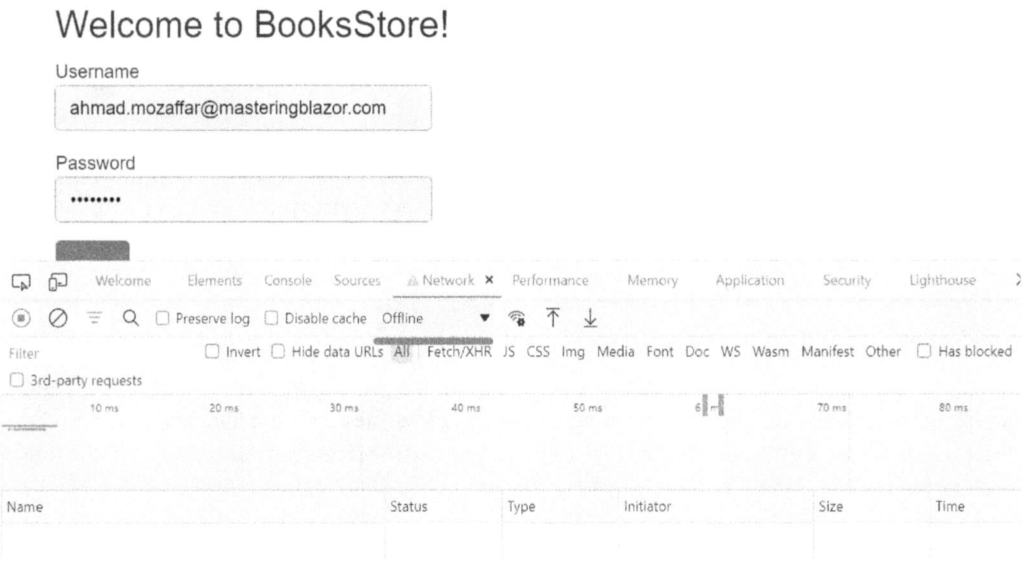

Figure 11.4 – Network throttling simulation in browser set to Offline

The `ErrorHandler` component solves critical issues and centralizes the important logic of our app. Next in this chapter, we will discuss the `ErrorBoundary` component, what it resolves, and when to use it.

Utilizing the ErrorBoundary component

In .NET 6.0, Blazor got a new component called `ErrorBoundary`. This component is very simple, easy to use, and powerful at the same time. The `ErrorBoundary` component handles any exceptions that occur during the rendering of a child component and displays a custom UI in place of the faulty component.

The `ErrorBoundary` component accepts two `RenderFragement` parameters: one called `ChildContent`, which represents the UI of the child components when no errors occur, and another called `ErrorContent`, which represents the content when an error occurs.

For example, when you use Twitter with a poor internet connection, you may see that the tweet details load successfully, but the replies below the tweet display an error message such as **Failed to load the replies** with a **Retry** button. This experience can be achieved with the ErrorBoundary component, but you need to keep in mind that the exception you need to handle with the ErrorBoundary error UI must not be handled inside the child component.

Let's see via an example how ErrorBoundary works. To demonstrate it, we will make an artificial exception with the **Counter** page. When the currentCount variable value becomes larger than 5, we will throw an exception, as shown here, in Counter.razor in the Pages folder:

```
. . .
private void IncrementCount()
{
  currentCount++;
  if (currentCount > 5)
    throw new Exception("Counter value must not exceed the value of
5");
    . . .
}
. . .
```

The **Counter** page is rendered within UserLayout, so we will place ErrorBoundary around the @Body property in the markup. So, open UserLayout.razor in the Shared folder. Add ErrorBoundary, as shown in the following code snippet:

```
. . .
<main class="px-5 py-4">
    <ErrorBoundary>
        <ChildContent>
            @Body
        </ChildContent>
        <ErrorContent>
            <h3>Counter must not exceed the value 5</h3>
        </ErrorContent>
    </ErrorBoundary>
</main>
. . .
```

Inside the component's <ChildContent>, we placed @Body to be rendered when everything is fine, and within the <ErrorContent> we put an <h3> with an error message.

Run the project, and you will see the **Index** page as normal, but if you navigate to the **Counter** page and click the **Increment Count** button six times, you will directly see an error, as shown in the following screenshot:

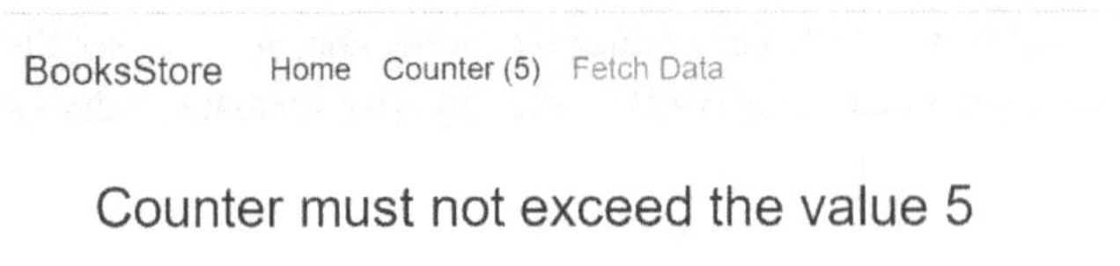

BooksStore   Home   Counter (5)   Fetch Data

Counter must not exceed the value 5

Figure 11.5 – ErrorBoundary rendering the error content

This little demo shows how useful `ErrorBoundary` is for scenarios such as the aforementioned Twitter example. Also, this demo explains why you should avoid using `ErrorBoundary` on a high-level scope, such as the layout, because if any exceptions happen in any component within any page, you will see the `Counter must not exceed the value 5` error. To get the best out of `ErrorBoundary`, you need to scope it for a specific component so it can render an appropriate and predictable error UI.

The exception details can be accessed within the `ErrorContent` tag via the `@context` variable if you want to use some data from the exception object properties in the rendered UI.

The `ErrorBoundary` component object has a method called `Recover`. This method resets the state of the `ErrorBoundary` component to a non-error state. To see how the `Recover` method works, add an `ErrorBoundary` variable to the `@code` section with `UserLayout`, and add a button below the `<h3>` error message that calls the `Recover` method. To populate the `_errorBoundry` variable with the component instance, assign it to the `@ref` directive as follows:

```
<ErrorBoundary @ref="_errorBoundry">
  ...
  <ErrorContent>
    <h3>Counter must not exceed the value 5</h3>
    <button class="btn btn-primary" @onclick="() => _errorBoundry.
Recover()">Reload</button>
  </ErrorContent>
</ErrorBoundary>
@code
{
  Private ErrorBoundary _errorBoundary;
}
```

If you test the same scenario again, when the error UI renders, you can click the **Reload** button, and the **Counter** page will appear again.

Overall, `ErrorBoundary` is a useful component when it's used correctly and in the right place. It's not very common to use it all over the place, but it can be very beneficial in some cases and helps with the user experience of the app.

So, at the end of this chapter, it's good to mention again that error handling doesn't have a magical standardized solution that resolves everything. Handling your APIs properly, having a good global error handler, and following the best practices while doing everything in the app can keep your app highly efficient and provide a satisfying experience for the end users.

Summary

After reading this chapter, you should be able to build an app that functions properly and delivers user value from the logic, UI, and reliability perspectives.

We started the chapter by talking about the importance of handling errors efficiently in your Blazor WebAssembly app. We started by going over how to handle the possible API responses, and then we implemented a global error handler that centralizes the processing of an exception. In the last section, we covered a cool component that's built for error handling from a UI perspective called `ErrorBoundary`. We demonstrated how we can use it and showed its capabilities.

After completing this chapter, you should be able to do the following:

- Understand the crucial need of handling error errors in single-page applications
- Handle failure requests when calling an API endpoint
- Build a centralized error handler that can be used across all the components
- Know how and when to use the `ErrorBoundary` component

In this next chapter, we will learn some techniques to follow to keep our app highly performant so the users not only enjoy an efficient and reliable app but will have a fast and lightweight experience too.

Further reading

- *Handle errors in ASP.NET Core Blazor apps*: `https://learn.microsoft.com/en-us/aspnet/core/blazor/fundamentals/handle-errors?view=aspnetcore-7.0#error-boundaries`

Part 3:
Optimization and Deployment

In this part, you will learn how Blazor works under the hood and how to take advantage of that to make your apps as efficient as possible, in addition to learning all the different techniques to make your app lighter and faster. After that, you will learn about component unit testing using bUnit and how to develop bulletproof components. Finally, you will learn about another Blazor WebAssembly hosting model, ASP.NET Core Hosted, and how to publish your app to the cloud.

This part has the following chapters:

11

Giving Your App a Speed Boost

Software that is fully functional is not always ready enough to ship. The app must open, let the user navigate between the pages, and update the data in the UI quickly enough to satisfy the user's expectations.

Blazor WebAssembly provides a set of features for optimizing the performance of the app, in addition to some tips that you can follow in the development phase that will increase the speed of your app significantly. This chapter covers some complex and specialized topics that are useful for creating apps where high performance and efficiency are the key. You may not encounter these situations often, but they are important to learn about so you are fully equipped and ready to handle any challenge that comes your way.

We will start this chapter by introducing the `Virtualize` component, in addition to some guidelines to help increase the efficiency of your Blazor components, especially when working on large-scale and data-intensive applications. We will also learn how to control and optimize the rendering process using the `ShouldRender` method, which can significantly improve app performance. Lastly, we will talk about Blazor WebAssembly's lazy loading feature, which reduces the initial size of the app and speeds up the load time.

This chapter will cover the following points:

- Increasing components' efficiency
- Rendering optimization with `ShouldRender`
- Decreasing the initial download size with assembly lazy loading

Technical requirements

The Blazor WebAssembly code and the API project used throughout this chapter are available on the book's GitHub repository in the *Chapter 11* folder:

```
https://github.com/PacktPublishing/Mastering-Blazor-WebAssembly/
tree/main/Chapter_11/
```

Make sure to have the API solution up and running beside the Blazor app.

Increasing components' efficiency

The first thing to do is to improve the speed of the component rendering process. Because, in Blazor WebAssembly, the rendering process happens fully in the browser, controlling the process of rendering can increase the overall speed of the app, especially in large, rich, and complicated UIs.

We will start this section by learning about the out-of-the-box `Virtualize` component. Then, we will learn some tips to keep in mind while developing Blazor components to maximize their efficiency.

Virtualize component

If you have a UI that renders a table with a large number of rows or a collection of UI components through a loop, the component will be slow and laggy because of the huge number of UI elements, especially while scrolling. One way to see this problem in action is to look at a chat or social media app. Imagine if the app loaded all 5,000 messages in a chatroom, or displayed all the posts in a newsfeed at once. That would create thousands of components in the UI, which would slow down the page and make it hard to use.

The `Virtualize` component introduced in Blazor in .NET 6.0 optimizes the rendering process for large lists of elements by only rendering the ones that are visible in the screen viewport. As the user scrolls, the `Virtualize` component dynamically updates the rendered elements based on the current viewport, removing the ones that are out of view. This way, the component and scrolling are efficient and smooth regardless of the number of elements in the UI.

The usage of the `Virtualize` component is straightforward; the following code snippet shows how to render the rows of a customer list that contain 10,000 objects the normal way:

```
<table class="table">
    <thead>
        <tr>
            <th>First Name</th>
            <th>Last Name</th>
            <th>Email</th>
        </tr>
    </thead>
    <tbody>
        @foreach (var customer in _customers)
        {
            <tr>
                <td>@customer.FirstName</td>
                <td>@customer.LastName</td>
                <td>@customer.Email</td>
```

```
            </tr>
        }
    </tbody>
</table>
```

The preceding code will result in rendering all the rows in the UI, which makes the component laggy. To use the `Virtualize` component, all you need to do is to replace the `foreach` loop with the `Virtualize` component:

```
...
<Virtualize Items="_customers">
    <tr>
        <td>@context.FirstName</td>
        <td>@context.LastName</td>
        <td>@context.Email</td>
    </tr>
</Virtualize>
...
```

The preceding code shows that the `Virtualize` component will iterate internally over the list items and render them in a performant fashion. At first, from a user perspective, it will be the same: a table of all the customer records. However, the difference will be clear when the user starts to scroll through the data; it will be smooth and efficient.

The `Virtualize` component takes four important parameters that provide more customization and performance gains:

- **ItemSize**: You can use this parameter to set the height of each item; this will increase the performance of the `Virtualize` component even more as the component won't need to do any calculations to know when to render the items near the viewport.

- **OverscanCount**: This property specifies how many items to render in advance when the user is near the edge of the visible area. This way, more items are ready to be shown when the user scrolls further. The main benefit of this property is that it reduces the flashing effect when scrolling fast, as the items are already rendered before the user sees them.

- **ItemsProvider**: This is a delegate that the `Virtualize` component triggers to fetch more items when the user scrolls; in most cases, this can be a method that fetches the data from an API.

- **Placeholder**: For items that are being fetched from the data source provided in the `ItemsProvider` parameter, `Placeholder` specifies a template to render in the meantime, until the item is fetched. This approach is common in modern apps such as social media platforms, within which, while scrolling down, more posts are fetched and you see an empty box fading in and out to let you know that the app is retrieving more content.

Structuring components for performance

It's quite tricky to manage the speed and the memory allocation of the code, and it requires some advanced skills to get it right. When it comes to the Blazor UI components, there are sets of practices and patterns that you can keep in mind while developing them.

It's good to know that each component is a fully independent object by itself, and it gets rendered in isolation from its parent and child components. Based on a test managed by the ASP.NET Core team (`https://learn.microsoft.com/en-us/aspnet/core/blazor/performance?view=aspnetcore-7.0#avoid-thousands-of-component-instances`), an average component that receives three parameters takes around 0.06 ms to render. The amount of time taken to render the component is tiny, but for complicated UIs that contain huge numbers of components to render, poor performance may appear.

Here are three tips that can improve the performance of components in UIs that contain hundreds and thousands of components to render:

- **Placing inline child components in their parent**: Generally, it's a good practice to have isolated components for code readability and maintenance. However, this may not be the best option for scenarios where performance is critical. For example, if you have a page that displays your store's orders in the last month as cards, and there are 1,500 orders in the UI, you don't need to create a separate component for the order card and render it inside a loop. Instead, you can have the card UI directly in the parent component. This can be translated into code as follows:

```
/* Instead of having it as a component */
@foreach (var order in _orders)
{
    <OrderCard OrderDetails="order" />
}


/* Consider having the OrderCard UI elements directly
   inside the foreach loop */
@foreach (var order in _orders)
{
    <div class="card">
        <h3>Number: @order.Number</h3>
        <p>Total: >
    </div>
}
```

The preceding code will save around 1,500 component rendering operations and instances in our scenario.

- **Avoiding many parameters when possible**: If the component needs to be nested, and nesting it directly in the parent component is not possible, avoid having flat parameters for each property. Every parameter received by a component is added to the total time the rendering operation takes. Combining parameters within a single object is going to help reduce the rendering time when thousands of component instances are being rendered.

The following code snippet shows an example of a customer card component that receives parameters that can be combined in a single object:

```
<h3>@DisplayName</h3>
@* Render more properties *@
...
@code {

    [Parameter]
    public int Id { get; set; }

    [Parameter]
    public string DisplayName { get; set; }

    /* More Customer properties */
}
```

Instead, you can combine the customer properties inside a class:

```
<h3>@Customer.DisplayName</h3>
@* Render more properties *@
...

@code {

    [Parameter]
    public CustomerDetails Customer { get; set; }

}
```

With all the simplicity of the preceding example, when rendering thousands of instances, each added parameter can add a massive amount of time to the overall rendering process.

- **Using fixed cascading parameters**: In the *Moving data between components* section in *Chapter 2, Components in Blazor*, we learned about cascading values and cascading parameters. By default, the IsFixed property is false for each cascading value, which means every recipient of the cascading value will have a subscription to track the changes to that value. When IsFixed is set to true, the recipients will receive the initial value of the cascading value but won't have a subscription to track future updates of it. Consider setting IsFixed to false whenever possible, as this will improve the performance, especially when the cascading value is received by huge numbers of recipients.

Optimizing the JavaScript calls

JavaScript interop from Blazor can also be a bit costly because JavaScript communication happens asynchronously by default. This is in addition to the JSON serialization and deserialization operations involved when passing parameters or returning function results between C# and JavaScript.

To avoid a bit of additional overhead when using JavaScript with your Blazor app, consider making JavaScript calls synchronously whenever possible. Like the example we gave in the *Turning an existing JS package into a reusable Blazor component* section in *Chapter 6, Consuming JavaScript in Blazor*, if you are using a service such as Google Analytics in your project, JavaScript calls to communicate with the **Google Analytics SDK** can be made synchronously without the need to wait until the call is finished.

Another optimization tip is to avoid repetitive JavaScript calls or calls within loops whenever possible. For instance, your app supports exporting invoices as PDFs, and there is a JavaScript function that exports an invoice object into a PDF on the client side that looks like this:

```
function exportInvoiceAsPdf(invoice) {
    // Logic to export the invoice as PDF
}
```

And there's a Blazor component that exports all the invoices in the table as PDFs:

```
foreach(var invoice in _invoices)
{
    await JS.InvokeVoidAsync("exportInvoiceAsPdf",
      invoice);
}
```

Instead, consider creating a new JavaScript method that accepts an array of invoices, iterates over the invoices, and calls exportInvoiceAsPdf:

```
function exportInvoicesCollectionAsPdf(invoices) {
    invoices.forEach(invoice => {
        exportInvoiceAsPdf(invoice);
    });
}
```

Now, in C#, you can make one JavaScript call instead of multiple asynchronous calls. We can also make it synchronously if the C# logic is not dependent on the result of this call:

```
JS.InvokeVoidAsync("exportInvoicesCollectionAsPdf ", invoices);
```

This little change can have a big impact on the performance of the export invoices feature.

Using System.Text.Json over other JSON packages

Either while consuming web APIs or storing objects in the browser storage, JSON is widely used in your Blazor WebAssembly apps. The process of serializing and reserializing JSON objects is a bit expensive in terms of performance, especially when it comes to large objects. `System.Text.Json` has been built from scratch by the .NET team to be the native library for dealing with JSON in your .NET apps. The main goal of the library is to improve the process of serialization and deserialization, as it's generally fast and efficient. So, the overall advice here is to use `System.Text.Json` over other packages when performance is a high priority, even though there are some trade-offs, as other libraries may be richer with more features.

Leveraging the aforementioned features and tips can noticeably accelerate your app's speed and efficiency. However, keep in mind that optimization is not always needed. For simple UIs or apps that deal with small amounts of data, having a deep level of optimization will decrease the development time and make the code a bit complicated, and the results won't be noticed. So, good optimization will give great results only if it's implemented at the right time.

Next, we will look at how Blazor re-renders the components in the hierarchy, and how we can prevent useless re-renders to keep the UI performant using the `ShouldRender` method, especially if we have hundreds and thousands of components within a page.

Rendering optimization with ShouldRender

In order to introduce the `ShouldRender` method and explain how to use it, first, we need to understand how parameters and events can affect the rendering of components in Blazor.

Blazor components exist in a hierarchy, with a root component that has child components, each child component can have its own child components, and so on. The re-render happens in the following scenario:

- When a component receives an event or a parameter changes, it re-renders itself and passes a new set of parameter values to its child components

- Each child component decides whether to re-render or not based on the type and value of the parameter values it receives:

  - If the parameter values are primitive types (such as `string`, `int`, `DateTime`, and `bool`) and they have not changed, the child component does not re-render

  - If the parameter values are non-primitive types (such as complex models, event callbacks, or `RenderFragment` values) or they have changed, the child component re-renders

- This process continues recursively down the component hierarchy until all affected components are re-rendered or skipped

This process of re-rendering is expensive, and it's very common too, as all the components are in the parent-child relationship. Blazor's ComponentBase class has a virtual method called ShouldRender and returns a bool value. By default, ShouldRender returns true, and the method gets called after OnParametersSet and before BuildRenderTree, which we learned about in *Chapter 1, Understanding the Anatomy of a Blazor WebAssembly Project*. If ShouldRender returns false, the component won't be re-rendered, so taking advantage of the ShouldRender method can prevent many useless renderings, keeping the app performant.

Let's see an actual example to understand how to use ShouldRender in a real-world scenario. On the **Index** page of the BooksStore project, we have the DataListView component, which renders the collection of books returned from the API as a BookCard component. The Index component passes a List<Book> object to DataListView, while DataListView passes a Book object for each BookCard component.

Initially, when the app runs, a welcome modal pops up on the screen. This triggers a change in the state of the **Index** page, and based on what we mentioned earlier in this section about the issue of re-rendering, the **Index** page will supply a new copy of the List<Book> object to DataListView, which will also supply a new copy of the Book object to each BookCard. So, whenever the welcome modal pops up or is closed, a change that's not related to the logic of the app causes DataListView and each BookCard to be re-rendered. The process is fast and unnoticeable because we have only 3 books, but if the API is returning 500 books, for example, every time you close the popup or click on the paging buttons that change the state of the **Index** page, you will notice that the app will lag. This is where ShouldRender comes into play. We know that once the books are retrieved from the API and each BookCard is rendered, the changes that affect the **Index** page shouldn't affect the rendering of the BookCard components.

Before we resolve the issue, let's see practically how each BookCard is being re-rendered every time the state of the **Index** page is changed.

Open the BookCard component in the Shared folder, override the OnAfterRender life cycle method, and add the following code:

```
. . .
protected override void OnAfterRender(bool firstRender)
{
    Console.WriteLine($"BookCard rendered for book
      '{Book.Title}'");
}
. . .
```

The OnAfterRender method will be triggered whenever the BookCard component gets rendered. Now, run the API and the Blazor project. When the page is rendered for the first time, you will see the following in the console window of **Developer Tools** in the browser:

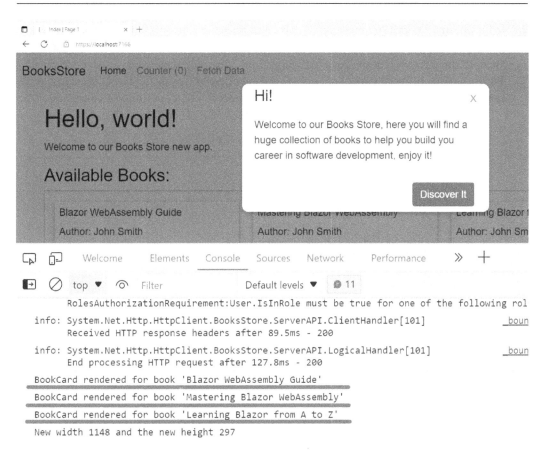

Figure 11.1 – BookCard OnAfterRender method gets triggered

That's expected, as it's the first render, but if you click the **X** button in the modal or the **Discover It** button, the state of the **Index** page will change, and you will notice the same messages will be printed again in the console window. That's because each BookCard gets rendered again.

The BookCard component won't change unless the Book parameter with its properties is changed; otherwise, the rendering result will stay the same during the component's lifetime. So, to prevent unnecessary re-rendering, we will do the following:

1. Declare two private fields in the component's @code section, one called _oldBookId of type string and another called _shouldRerender of type bool:

   ```
   ...
   private string _oldBookId = string.Empty;
   private bool _shouldRerender = false;
   ...
   ```

2. In the OnParametersSet method, which is called when the component receives new parameter values from its parent component, we compare the current Book.Id with _oldBookId. If they are different, it means that the book has changed, and the component needs to be re-rendered. We update _oldBookId and set _shouldRerender to true, and if the IDs match, we can skip the re-rendering, so we set _shouldRerender to false in the else block:

```
. . .
protected override void OnParametersSet()
{
    if (Book == null)
        throw new ArgumentNullException(nameof(Book));
    if (Book.Id.Equals(_oldBookId))
    {
        _oldBookId = Book.Id;
        _shouldRerender = true;
    }
    else
    {
        _shouldRerender = false;
    }
}
. . .
```

3. Finally, we can override the ShouldRender method and let it return the _shouldRerender value, which will be false most of the time, so the component won't be re-rendered:

```
. . .
protected override bool ShouldRender()
{
    return _shouldRerender;
}
. . .
```

Now, if you run the project again and monitor the console window, you will notice that the messages being written inside OnAfterRender will be printed only at the first rendering; then, re-rendering will be skipped.

> **Note**
>
> You shouldn't always override ShouldRender and prevent re-rendering. This approach should only be taken when your component UI doesn't change after the first render, regardless of the parameter values. Another example where ShouldRender is needed is while adding or manipulating items within a list that you don't want to be rendered until the full list is ready.

In the next chapter, we will learn how Blazor renders the components internally with `RenderTree` so we can optimize even more whenever we need to.

Next, we will cover a powerful Blazor WebAssembly feature that can shrink the initial load size of the app, which makes it load faster when it opens. This will give us fully optimized software not only during runtime but also during initialization.

Decreasing the initial download size with assembly lazy loading

In single-page applications in general and Blazor WebAssembly specifically, the full app packages, scripts, and stylesheets are downloaded for the first time to the browser so the app is fully functional on the client side. The average Blazor WebAssembly app has a size of 2 MB when it is published. That's not very big, but when the app gets more complicated and more packages and assemblies are involved, the size can increase significantly.

In this section, we will learn about **Blazor WebAssembly's lazy loading** feature, which was released in .NET 6.0 to help reduce the initial download size to make first-time loading faster. Additionally, Blazor provides some out-of-the-box features, such as runtime relinking, compression, and trimming, that decrease the app size, but we will cover those features in the *Blazor WebAssembly app prerelease final checks* section *Chapter 14, Publishing Blazor WebAssembly Apps*.

By default, when you run the app for the first time, all the assemblies that are used are downloaded. For huge apps with large assemblies, that could decrease the startup time until all the assemblies are ready. In .NET 6.0, Blazor introduced the lazy loading feature, which waits for some defined assemblies to load until they are required instead of loading them at startup time.

To use the lazy loading feature effectively, you should organize your solution with Razor Class Library projects when possible. For example, suppose you have an e-commerce app with a statistics page that shows charts of your sales. You can put the chart components in a separate Razor Class Library project and mark it for lazy loading. This way, the chart assembly will only be loaded when you navigate to the statistics page, not when you launch the app. This can improve the app's startup performance and save bandwidth, especially if the statistics page is rarely accessed or restricted to certain users.

Let's see a practical example of lazy loading an assembly in our `BooksStore` project. In addition to the main Blazor WebAssembly project, we have a Razor Class Library project called `BooksStore.Blazor.Components`. This project contains reusable components mostly used in the **BookForm** page, the page the admin uses to add new books to the system. We can mark the `BooksStore.Blazor.Components` package to be lazily loaded when the admin navigates to `/book/form` instead of loading it at first.

Let's see how the assemblies are being loaded first before we start achieving the desired outcome. Run the Blazor project and open a new private window in your browser, as all the DLLs and other app files are cached in the normal browser instance that you use for testing. Open the **Network** tab in **Developer**

Tools in the browser, then navigate to the **Index** page of the app (`https://localhost:7166`). You will notice all the JavaScript files and DLLs are loaded, including `BooksStore.Blazor.Components`, as shown here:

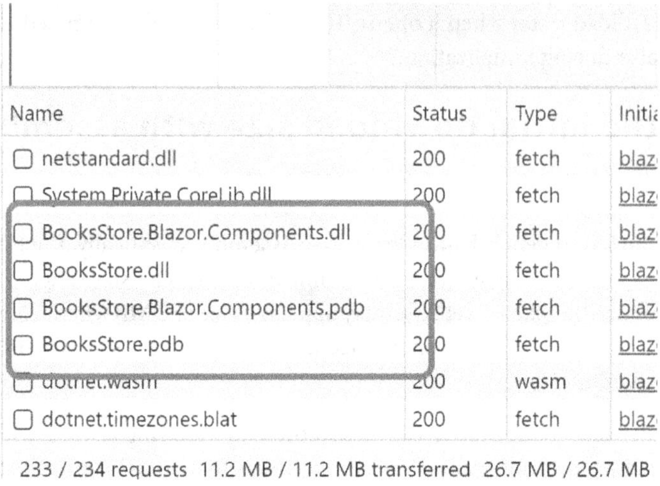

Name	Status	Type	Initia
☐ netstandard.dll	200	fetch	blaz
☐ System.Private.CoreLib.dll	200	fetch	blaz
☐ BooksStore.Blazor.Components.dll	200	fetch	blaz
☐ BooksStore.dll	200	fetch	blaz
☐ BooksStore.Blazor.Components.pdb	200	fetch	blaz
☐ BooksStore.pdb	200	fetch	blaz
☐ dotnet.wasm	200	wasm	blaz
☐ dotnet.timezones.blat	200	fetch	blaz

233 / 234 requests 11.2 MB / 11.2 MB transferred 26.7 MB / 26.7 MB

Figure 11.2 – The files and assemblies loaded for the first time

Now, let's get started with implementing assembly lazy loading for `BooksStore.Blazor.Components` so that this package is installed when we navigate to the `/book/form` page:

1. Set `BooksStore.Blazor.Components` to be loaded when navigating to the `/book/form` page.

2. Open the `BooksStore.csproj` file in the Blazor project and add the following snippet, which will mark `BooksStore.Blazor.Components.dll` to be lazily loaded:

```
...
<ItemGroup>
    <BlazorWebAssemblyLazyLoad
      Include="BooksStore.Blazor.Components.dll" />
</ItemGroup>
...
```

By just adding this, `BooksStore.Blazor.Components` won't be loaded at the first load of the app anymore. So, next, we need to modify `App.razor` so it will load `BooksStore.Blazor.Components` when navigating to the `/book/form` page.

3. In `App.razor`, add the following namespaces and inject `LazyAssemblyLoader` and `ILogger<App>`. `LazyAssemblyLoader` is the service that's responsible for loading the desired assemblies:

```
. . .
@using Microsoft.AspNetCore.Components.Routing
@using Microsoft.AspNetCore
   .Components.WebAssembly.Services
@using Microsoft.Extensions.Logging
@using System.Reflection;
@inject LazyAssemblyLoader AssemblyLoader
@inject ILogger<App> Logger
. . .
```

4. Add a `@code` section and define a `List<Assembly>` instance:

```
. . .
@code
{
    private List<Assembly> _lazyLoadedAssemblies =
        new();
}
```

5. Define a method that takes a `NavigationContext` parameter. This parameter contains information about the current navigation. The method will be assigned to an `EventCallback` named `OnNavigateAsync` in the `Router` component we learned about in the *Understanding routers and pages* section in *Chapter 4, Navigation and Routing*. The method will check whether the user has navigated to `/book/form`. If so, it will load the assembly from `BooksStore.Blazor.Components.dll` and add it to `_lazyLoadedAssembliesList`:

```
. . .
private async Task OnNavigateAsync(NavigationContext
   args)
{
    try
    {
        if (args.Path.Equals("book/form",
          StringComparison.InvariantCultureIgnoreCase))
        {
            var assemblies = await AssemblyLoader
              .LoadAssembliesAsync(
                    new[] { "BooksStore
                       .Blazor.Components.dll" });
                _lazyLoadedAssemblies
                  .AddRange(assemblies);
```

```
                }
            }
            catch (Exception ex)
            {
                Logger.LogError("Error: {Message}",
                  ex.Message);
            }
        }
        ...
```

6. In the markup section, assign the previous method to OnNavigateAsync EventCallback of the Router component, and assign the _lazyLoadedAssemblies list to the AdditionalAssemblies parameter of the Router component too, as shown here:

    ```
    ...
    <Router AppAssembly="@typeof(App).
    Assembly"    AdditionalAssemblies="_
    lazyLoadedAssemblies"    OnNavigateAsync="OnNavigateAsync">
    ...
    ```

    OnNavigateAsync will be triggered when the user navigates between pages, so, based on the written logic, when the user navigates to /book/form, BooksStore.Blazor. Components.dll should be loaded.

Now, let's make sure that what we have done is working properly. Let's run the project again, open it in a private window one more time, and monitor the **Network** tab. The first thing we should notice is that BooksStore.Blazor.Components is no longer loaded when we open the app for the first time:

Name	Status	Type	Initiator	Size
mscorlib.dll	200	fetch	blazor.webasse…	23.1 kE
netstandard.dll	200	fetch	blazor.webasse…	33.4 kE
System.Private.CoreLib.dll	200	fetch	blazor.webasse…	1.3 ME
BooksStore.dll	200	fetch	blazor.webasse…	35.4 kE
BooksStore.pdb	200	fetch	blazor.webasse…	49.3 kE
dotnet.wasm	200	wasm	blazor.webasse…	1.0 ME
dotnet.timezones.blat	200	fetch	blazor.webasse…	75.3 kE

Figure 11.3 – BooksStore.Blazor.Components not loaded at first launch

Then, log in with the admin account so the **Add Book** button appears in the menu. Click on the **Add Book** button, which will take you to /book/form. You will notice in the **Network** tab that the BooksStore.Blazor.Components.dll file is loaded the moment you navigate to that page:

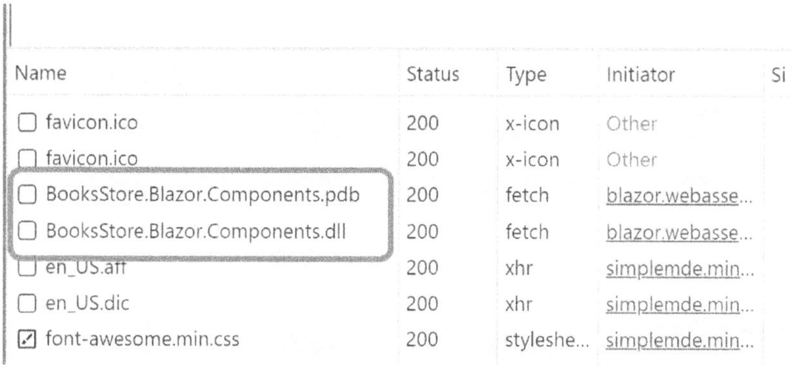

Figure 11.4 – BooksStore.Blazor.Components is loaded when navigating to /book/form

> **Note**
>
> You can mark multiple assemblies to be lazily loaded, and you can load multiple assemblies at the same time. The `LoadAssembliesAsync` method accepts a collection of assembly names to load, but for the purpose of our example, we need only to lazily load one assembly.

Assembly lazy loading is a powerful feature and has a clear impact on the loading time and size of an app, especially when you have an enterprise-level application with many large assemblies. One more reminder: defining a good separation in your solution's projects is a crucial factor in being able to utilize the assembly lazy loading feature. For example, components that are not always accessed, are quite large, and reference multiple packages should be placed into a different project so you can mark them to be lazily loaded.

Summary

This chapter discussed the importance of optimizing the performance of a software application beyond ensuring that it is fully functional. The focus was on Blazor WebAssembly and the various features it provides to help improve app performance. The chapter covered three main areas: increasing component efficiency, rendering optimization with `ShouldRender`, and decreasing app size with lazy loading. The `Virtualize` component was introduced, along with guidelines for improving Blazor components' efficiency, especially in data-intensive applications. The `ShouldRender` method was also discussed. We highlighted its importance in controlling and optimizing the rendering process to improve app performance. Finally, the chapter explored Blazor's lazy loading feature, which reduces the app's initial size and speeds up the load time.

After completing this chapter, you should be able to do the following:

- Understand the importance of performance in Blazor WebAssembly apps
- Use the `Virtualize` component to render a large number of UI components efficiently

- Use guidelines to keep your component optimized and performant in rich and intensive UIs

- Understand rendering in the component hierarchy and avoid useless rendering with `ShouldRender`

- Reduce the app size and improve the load time with Blazor lazy loading

In the next chapter, we will explore Blazor's `RenderTree` and examine how it updates the UI. Through practical examples, readers will gain a clear understanding of how Blazor works underneath the hood.

Further reading

- ASP.NET Core Blazor performance best practices: `https://learn.microsoft.com/en-us/aspnet/core/blazor/performance?view=aspnetcore-7.0`

12

RenderTree in Blazor

We have seen how we develop components, manipulate the UI, build forms, and optimize our components and project for faster and more efficient performance. It's now time to go deep and learn how Razor components are compiled and how Blazor manipulates the UI you see in the browser.

We will start by learning a bit more about the difference between **Single-Page Applications** (**SPAs**) and traditional web apps in terms of how the UI works, then we will go deeper by defining **RenderTree** in Blazor and what happens to .razor files at compilation. After covering the theory, we will develop a simple component using purely C# and learn about how to take advantage of this advanced scenario, helping us build more performant apps.

This chapter covers some advanced topics that may not be directly relevant for practical applications of Blazor and day-to-day work, but they can help you gain a deeper understanding of how Blazor works under the hood and go through scenarios that require a high level of optimization. If you are curious about the inner workings of Blazor, this chapter is for you. This chapter will cover the following:

- SPAs versus traditional web apps with rendering
- What is RenderTree in Blazor?
- Building a component with RenderTree
- Controlling the rendering using the @key directive

Technical requirements

This chapter uses a new Blazor WebAssembly project called RenderTreeSample and not the original BooksStore project used in other chapters.

The code used is available in the book's GitHub repository:

https://github.com/PacktPublishing/Mastering-Blazor-WebAssembly/tree/main/Chapter_12

How rendering happens in SPAs

What you see in any web app, in the UI, is the result of the rendering process of the HTML code that represents the page or components.

In traditional web apps, when the browser requests the page, the server either sends a full HTML page in the case of static web apps, or, in situations of apps such as ASP.NET MVC, it constructs a full HTML string after doing data and UI manipulation, and then sends it to the client. The browser takes that HTML and renders it as is.

On the other hand, in the case of SPAs, the browser initially sends a request and receives a simple HTML document alongside the **JavaScript (JS)** libraries, or the DLLs if the framework is Blazor. The browser renders that simple page, then the logic of the JS or .NET library starts to build HTML pieces and injects or replaces them in the UI. This all happens on the browser side through a process known as manipulating the **Document Object Model (DOM)**.

What is the DOM?

Basically, when the browser renders a piece of HTML, it builds a programmatic representation for that document in a tree-structure manner. Each node of that tree represents an object inside the document. This programmatic interface wraps all the document's elements, content, and attributes.

The DOM is what is used to manipulate the UI of the application. Let's see an example in the following HTML document:

```
<!DOCTYPE html>
<html lang="en">
<head>
    <title>Title</title>
</head>
<body>
    <h1>Welcome message</h1>
    <button>Update Message</button>
</body>
</html>
```

The previous HTML represents a full document, and its DOM can be represented as follows:

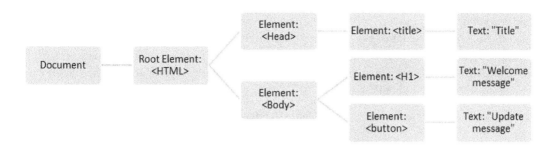

Figure 12.1 – DOM for the HTML document

Interactive apps provide a rich UI that's updated correspondingly in a quick and performant fashion. For example, when adding a new data item (a book, for example), the UI reflects the changes immediately, instead of requesting the full page from the server by making a full reload.

Updating the DOM using JS

In the previous example of the message and the button, very basically, to update the text inside the <h1> tag, we will add a little JS script to update the content when clicking on the button. The following script added at the end of <body> will perform the update by accessing the <h1> tag from the DOM and performing the message update:

```
...
<button onclick="updateMessage()">Update Message</button>
    <script>
        function updateMessage() {
            var firstElement = document.body.children[0];
            firstElement.innerHTML = 'Updated message';
        }
    </script>
</body>
```

In the snippet, we have referenced the <h1> tag as it's the first child within the body of the document; then, we used its innerHTML property to update the content.

If you open the HTML document and click the **Update message** button, you will notice that the message gets updated. This is a very basic example, but it highlights the core principle of advanced web apps and SPAs.

Of course, to reference elements, you can use more complex selectors, such as using the ID of the tag or the classes applied to it, but this example just showed how we can manipulate the DOM in a basic manner.

While you are using any modern social media platform, you'll notice how fast it is in updating the UI despite how complicated it is. When a new post is received, it directly pops up in the UI without the need to update the full page or refresh it. In complex apps built with modern frameworks such as ReactJS, Angular, or Blazor, manipulating the DOM as efficiently as possible is the key factor in choosing the framework and building the UI.

In the next section, we will see how Blazor updates the DOM internally using what is called `RenderTree`.

What is RenderTree in Blazor?

While developing components in Blazor and using some bindings, we don't update the DOM directly. Between our components and the DOM, Blazor creates an in-memory programmatic lightweight object called `RenderTree` that represents the current DOM's state.

Due to the nature of `RenderTree` as a C# object, it's easy to manipulate its state and content instead of manipulating the DOM directly, because in SPAs, the process of manipulating the DOM is heavy and complex. When a change occurs in the app and it requires the UI to be updated, Blazor updates the state of `RenderTree` and then compares the updated status with the original one using an advanced *diffing* algorithm. The algorithm is responsible for identifying the differences between the original state and the new one in an efficient manner. This all happens in `RenderTree`, rather than in the DOM directly, taking only the differences and applying only what has changed to the DOM, which leads to a very performant operation.

The great thing about SPAs is that you can have heavy client-side logic and operations and then they are outputted on the UI, so instead of reflecting the changes that happen in the logic directly to the UI, it finalizes the operation, checks for changes that happened to the UI, and finally, applies them. In the *Controlling the rendering using the @key attribute* section later in this chapter, we will see with an example how `RenderTree` updates the DOM very efficiently, but for now, let's see how it's structured.

> **Note**
> Blazor doesn't re-render the full element or component every time. If you update text within a `<p>` tag, only the text inside the tag gets updated; everything else remains as is through a sufficient process or operation.

Understanding the structure of the component in RenderTree

When you build any component in Blazor using a `.razor` file, basically, at compile time, this Razor file turns into a normal C# class and the markup gets translated into `RenderTree`. Every component inherits from the `ComponentBase` class, which contains a method called `BuildRenderTree` and is supplied with a parameter of the `RenderTreeBuilder` type, which is used to build the `RenderTree` instance of the component.

If we check the compiled code of the default `Counter.razor` page of any Blazor project, we will see the following class:

```
using Microsoft.AspNetCore.Components;
using Microsoft.AspNetCore.Components.Rendering;
using Microsoft.AspNetCore.Components.Web;
namespace RenderTreesSample.Pages
{
    [RouteAttribute("/counter")]
    public partial class Counter : ComponentBase
    {
        protected override void
            BuildRenderTree(RenderTreeBuilder __builder)
        {
            __builder.OpenComponent<PageTitle>(0);
            __builder.AddAttribute(1, "ChildContent",
                (RenderFragment)((__builder2) =>
            {
                __builder2.AddContent(2, "Counter");
            }
            ));
            __builder.CloseComponent();
            __builder.AddMarkupContent(3, "\r\n\r\n");
            __builder.AddMarkupContent(4,
                "<h1>Counter</h1>\r\n\r\n");
            __builder.OpenElement(5, "p");
            __builder.AddAttribute(6, "role",
                "status");
            __builder.AddContent(7, "Current count: ");

            __builder.AddContent(8, currentCount);

            __builder.CloseElement();
            __builder.AddMarkupContent(9, "\r\n\r\n");
            __builder.OpenElement(10, "button");
            __builder.AddAttribute(11, "class",
                "btn btn-primary");
            __builder.AddAttribute(12, "onclick",
                global::Microsoft.AspNetCore
                .Components.EventCallback.Factory
                .Create<MouseEventArgs>(this,
                IncrementCount));
            __builder.AddContent(13, "Click me");
            __builder.CloseElement();
        }
```

```
            private int currentCount = 0;
            private void IncrementCount()
            {
                currentCount++;
            }
        }
    }
```

As you can see, the same `Counter.razor` file that has a little piece of markup and a small, one-line C# method, has turned into a giant class. That's the magic of Blazor. If you notice the `BuildRenderTree` method, it constructs the same component content of the `.razor` file created by the default Blazor project template but using pure C# objects and methods.

Note

To be able to see the generated classes from `.razor` files, make sure to add `<EmitCompilerGeneratedFiles>true</EmitCompilerGeneratedFiles>` `<CompilerGeneratedFilesOutputPath>.</CompilerGenerated FilesOutputPath>` to the `.csproj` file of your Blazor app inside the `<PropertyGroup>` tag. Then, when you build the project, you can find the compiled files in the following path:

`Project_Root_Folder/ Microsoft.NET.Sdk.Razor.SourceGenerators/ Microsoft.NET.Sdk.Razor.SourceGenerators.RazorSourceGenerator`

Also, make sure to disable them as they will affect the upcoming build operations.

Understanding RenderTree's sequence numbering of elements

A very important thing to notice in every method that is added to `RenderTree`, such as `OpenComponent`, `AddAttribute`, and `AddMarkupContent`, is that they all receive a static number as the first parameter and that's the sequential number for every node in the tree. This is the number being used by the `RenderTree` diffing algorithm to determine the differences when the state changes. The number must be sequential and it's preferable for it to be a static raw number, not referenced from a variable or calculated inside a loop.

Because RenderTree's goal is to minimize the changes to be made to the DOM as much as possible, the sequential number is an important part of the process. The following example shows how sequential numbering can help with making the process more efficient:

1. Modify the `Counter` component by adding a `bool` variable called `showCounter` to the `@code` section, and a method to toggle the value of that variable, as shown:

    ```
    ...
    private bool showCounter = true;
    private void ToggleCounter()
    {
    ```

```
        // Toggle the value of the showCounter variable
        showCounter = !showCounter;
    }
    ...
```

2. Wrap the `<p>` tag that renders the `currentCounter` variable in an `if` condition to hide it if `showCounter` is set to `false`:

```
    ...
    @if (showCounter)
    {
        <p role="status">Current count: @currentCount</p>
    }
    ...
```

3. Add a button after the **Click me** button to toggle the counter visibility:

```
    ...
    <button class="btn btn-danger"
      @onclick="ToggleCounter">Show/Hide Counter</button>
    ...
```

After running the app and navigating to the `/counter` page, you should see the following:

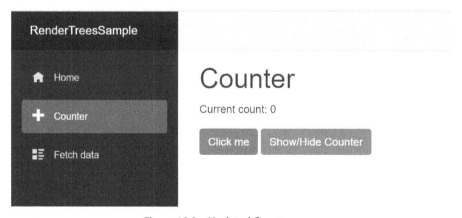

Figure 12.2 – Updated Counter page

Let's suppose that `RenderTree` builds the components and gives the `<p>` tag the sequential number 2 and the **Click me** button the number 3, as shown in the following table:

Sequential Number	Element
2	`<p>` tag that wraps the `currentCounter` value
3	`Click me <button>` tag

Table 12.1 – The settings of the RenderTree state

When the user clicks the **Show/Hide Counter** button, the `<p>` tag will be hidden, so the Blazor app must take the following step only to update the DOM: remove the element with sequence number 2.

If we are building the same component manually, as we will do in the next section, instead of using a fixed sequence number, we will use a variable and its value will increase by the number of components in the UI. In such a case, if we suppose that the `<p>` tag is the second element, as shown in *Table 8.1*, and the sequence number is stored in a variable called `sequenceNumber`, we will have the following:

Sequential Number	Element
`sequenceNumber + 2`	`<p>` tag that wraps the `currentCounter` value
`sequenceNumber + 3`	`Click me <button>` tag

Table 12.2 – The settings of the RenderTree state with variable sequence numbers

After the user clicks **Show/Hide Counter** to hide the counter from the UI, `RenderTree` has to do additional calculations to detect the changes, and after `<p>` gets removed, it has to give new sequence numbers to all the components after it.

With that, you should have a good idea of how things are going under the hood. Let's now try to build a basic component using `RenderTree` directly.

Building a component with RenderTree

Building components can be done using just a `.razor` file, which is the recommended approach most of the time, but it's good to learn how to also build them using `RenderTree`, to learn how things work under the hood. This will help you build better components with Razor. For example, with `RenderTree` you can control how Blazor determines the changes to render in the UI, which can improve performance in some special cases. We will see an example of this in the *Controlling the rendering using the @key directive* section.

It's rare to encounter a scenario where you're required to use the `RenderTree` method, and it's also not recommended, as building complex components is extremely hard with `RenderTree` compared to `.razor` files.

To keep things simple, we will build the same little UI with the <h1> tag and <button> and configure it to update its content when we click on the button as we did in the *Updating the DOM using JS* subsection earlier in the chapter. The result in Razor should look like the following:

```
@page "/render-tree-demo"
<h1>Welcome </h1>
<button class="btn btn-primary"
  @onclick="UpdateName">Update Name</button>

@code {
    private string name = "Blazor";
    private void UpdateName()
    {
        name = ".NET";
    }
}
```

Let's achieve the same thing but with RenderTree:

1. In the Pages folder, create a new C# class file, instead of a .razor file, and call it MessageViewer.cs.

2. The Razor component inherits from the ComponentBase class, so let's implement that and add the [Route] attribute, which is equivalent to the @page directive in Razor to set a route value for the page:

    ```
    [Route("message-viewer")]
    public class MessageViewer : ComponentBase
    {
    }
    ```

3. Inside the class, we need to declare a string variable and a method called UpdateMessage to update its content:

    ```
    . . .
    private string _message = "Original message";
    private void UpdateMessage()
    {
        _message = "Updated message";
    }
    . . .
    ```

4. Override the `BuildRenderTree` method that we will use to build the component:

    ```
    protected override void
      BuildRenderTree(RenderTreeBuilder builder)
    {

    }
    ```

5. Now, we can use the builder object and its methods to build the markup, as we have seen earlier in the `Counter` component. We will start by opening an `<h1>` tag with fixed text content inside, adding further HTML content with the _message variable, and finally, closing the tag:

    ```
    builder.OpenElement(0, "h1"); // Used to open an HTML
      element
    builder.AddContent(1, "Message: "); // Add content
      inside the HTML element
    builder.AddContent(2, _message);
    builder.CloseElement(); // Used to close an open
      element
    ```

    The previous code will render the equivalent of the following code in Razor:

    ```
    <h1> Message: @_message</h1>
    ```

    Create the button and add its class and `@onclick` attributes. The value of `@onclick` is an `EventCallback` instance and we can use the `EventCallback.Factory.Create` method to assign the value:

    ```
    . . .
    builder.OpenElement(3, "button");
    builder.AddAttribute(4, "class", "btn btn-primary");
      // Add attribute with its name and value
    builder.AddAttribute(5, "onclick",
      EventCallback.Factory.Create(this, UpdateMessage));
      // Used to set the value of the @onclick to the
      UpdateMessage method
    builder.AddContent(6, "Update message"); // Add the
      text of the button
    builder.CloseElement();
    . . .
    ```

Our component is now ready, but if you look at the original Razor code and the final C# code we have written, you will notice how complicated it is to build with `RenderTreeBuilder`. It's very easy to make mistakes in the tags, sequence number, and so on, which will lead to unlikely behaviors.

To see the result, run the project and navigate to `/message-viewer`. You should have just the following component:

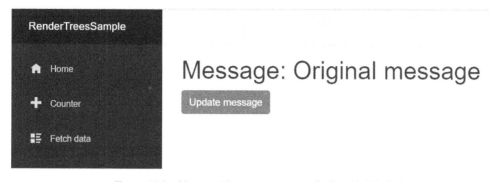

Figure 12.3 – MessageViewer component built with RenderTree

If you click the **Update message** button, you will notice the text changes to the updated message.

Controlling the rendering using the @key directive

When it comes to rendering collections by iterating over the items and rendering some UI elements, the powerful mechanism of Blazor `RenderTree` and its diffing algorithm will observe the changes happening to the list (insert, update, and delete) and update the UI accordingly. Blazor assigning and indexing for each item gets rendered in the UI based on the items in the list at runtime. That's the normal approach for most cases.

Because the sequence numbers are assigned at runtime, Blazor will start sequentially from the first item in the list to the last. So, let's create a sample component that will simulate the scenario, as follows:

1. In the Pages folder, create a Razor component and call it `key-directive-sample.razor`.

2. Write the following code inside it:

```
@page "/key-directive-sample"
<h3>Key Attribute Sample</h3>
<ul>
    @foreach (var item in numbers)
    {
        <li>@item</li>
    }
</ul>
<button @onclick="Insert">Sort</button>
@code {
    private int counter = 4;
    private List<int> numbers = new List<int>
    {
        0, 1, 2, 3
    };
    private void Insert()
```

```
    {
        numbers.Insert(0, counter++);
    }
}
```

If you run the project and navigate to /key-directive-sample, you will notice just a bullet list of numbers with a button. If you click on the button, a new number will be inserted at the top of the list:

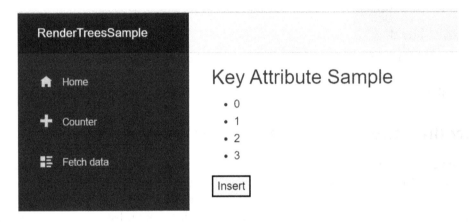

Figure 12.4 – KeyAttributeSample component's initial state

Blazor will assign a sequence number for each <li> element in the loop initially and for each new item being inserted. The following table shows the sequence numbers for the items based on the 0 index for the sample:

Sequential Number	Element
0	<li>0</li>
1	<li>1</li>
2	<li>2</li>
3	<li>3</li>

Table 12.3 – The sequence numbers of the numbers list in the UI

That works fine, but if we insert a new item at the top of the list, how will `RenderTree` know to preserve sequence number 0 for element `<li>0</li>`? If we click the **Insert** button, number 4 will be added to the top and Blazor will give the new element a new sequence number starting from 0. That means all the existing items will be pushed down in terms of sequence number, and `RenderTree` will lose the tracking of the initial status and will end up rerendering the list, because the elements will have the following new sequence numbers:

Sequential Number	Element
0	`<li>4</li>`
1	`<li>0</li>`
2	`<li>1</li>`
3	`<li>2</li>`
4	`<li>3</li>`

Table 12.4 – The sequence numbers of the numbers list after inserting a number

Despite the expected results showing up in the UI, the approach used was not efficient for the following reasons:

- Blazor must assign a new sequence number and will lose track of the original state for some elements, or maybe all of them in the case of inserting a number at the top of the list. That leads to rerendering and affects the performance of the app.

- If you have text input elements for each item, due to the rerendering, the cursor focus will be lost, and the state will be back to its original state which leads to unlikely behaviors.

To resolve that, we need a mechanism to tell `RenderTree` explicitly what the key is of each rendered item, so when any modification happens to the data, `RenderTree` knows exactly what elements were affected and it only updates them in the DOM.

In the previous example, let us assume `RenderTree` knows how to map each `<li>` tag for a specific known key instead of giving it a new one automatically. When an item is added to the beginning of the list, it knows that none of the other items have changed, so it only applies element insertion at the top, which is very efficient and keeps the state of the unmodified items.

Blazor provides us with an attribute called **@key**. We can use it to control the mapping process of `RenderTree`, especially within the loops. `RenderTree` will use the supplied value as a key for the element to be used in the diffing algorithm and detect the changes.

To improve the previous example, all we need to do is to add the @key value for the <li> element inside foreach, and give it the value of the number being added, as shown:

```
...
@foreach (var item in numbers)
{
    <li @key="item">@item</li>
}
...
```

The value of @key must be unique; otherwise, the app will throw an exception. So, you can use either the object itself you are iterating, or use the Id property of the object. In the case of rendering a list of books, for example, you can use the ID of the book as the value for it.

> **Note**
>
> You should use @key when you render a list in the UI that can change, for example by adding or removing items. However, when you render a static list or a list that only grows at the end, you don't need @key because Blazor will keep the sequence numbers intact. To learn more about how and when to use it in more detail, check out the following link: https://learn.microsoft.com/en-us/aspnet/core/blazor/components/?view=aspnetcore-7.0#use-key-to-control-the-preservation-of-elements-and-components.

With that, you have learned more about Blazor and how great, efficient, and clever this framework is.

Summary

In this chapter, we went over how Blazor handles the DOM manipulation process using an abstraction layer called RenderTree between the DOM and our components. We went over how things work inside Blazor, then built a component using purely C# with RenderTreeBuilder. Finally, we saw a new directive called @key that helps us have a more efficient rendering process and more predictable output when rendering collections.

So, after building components in the previous chapter, this one has completed the picture and has fed our curiosity. But the most important takeaways of this chapter are as follows:

- Blazor uses RenderTree to track changes in the UI, aggregate them, and apply them to the DOM efficiently.

- RenderTree doesn't rerender the markup every time an update is made. It only updates what has changed in the UI.

- Use @key whenever you render a list of items, for example, within a foreach loop, to have a better-controlled rendering process.

The next chapter will teach us how to write unit tests for both the UI components and the app logic, and why testing is important for making our components reliable and robust.

Further reading

- What is the DOM? `https://en.wikipedia.org/wiki/Document_Object_Model`
- More about RenderTrees: `https://blazor-university.com/components/render-trees/`

13

Testing Blazor WebAssembly Apps

Testing is an essential process in software development that helps ensure that the application functions as expected and meets the desired quality standards. In this chapter, we will briefly introduce the concept of testing software, the different types of testing, and their importance, and then we will take a deep dive into testing Blazor WebAssembly apps.

This chapter will introduce you to the **bUnit** library, which is used for writing **unit tests** for Blazor components, and then we will start writing some unit tests that cover different scenarios and different types of components. We will learn advanced concepts in unit testing, such as mocking and faking using **Moq**, and we will cover the built-in services in bUnit in addition to writing unit tests that cover those features. Lastly, we will briefly introduce the **Playwright** library and **end-to-end** (E2E) **testing** in Blazor WebAssembly.

This chapter will cover the following topics:

- Testing Blazor components overview
- Getting started with bUnit
- Writing components unit tests with bUnit
- Mocking and faking in unit tests with Moq and bUnit
- Introducing Playwright for E2E tests

Technical requirements

The Blazor WebAssembly code and the API project used throughout this chapter are available in the book's GitHub repository in the folder for *Chapter 13*:

```
https://github.com/PacktPublishing/Mastering-Blazor-WebAssembly/
tree/main/Chapter_13/
```

Testing Blazor components overview

Testing software is a crucial part of the development process, not only in SPAs and Blazor WebAssembly but in all elements of software. *Testing* as a term has multiple meanings and approaches in the software industry and long books could be written about it, but we will focus on component testing developed in Blazor WebAssembly. We will examine the two testing types: unit testing and E2E testing.

For every feature we have developed so far, if we run the project, navigate to the target page or component, and use it to see if it will behave as expected, we are only conducting one type of testing. This type of testing is called end user testing, which is not reliable. To ensure the app logic, behavior, and components are working as expected, tests should be written at the code level, so more common and rare scenarios can be simulated.

The first type of testing we will examine is called **unit testing**, which is a fundamental type of testing in software development. It involves testing every unit of the system, such as logic, behavior, or a component, in isolation from all its dependencies and environment to ensure it is functioning as expected.

There is a misunderstanding about unit testing. I have seen many developers ignore writing tests as this process requires more code, which slows the development process. However, during the day-to-day development process, unit testing accelerates the development time in addition to making the components more reliable. This seems weird at first; how does writing more code make the process faster? But, basically, writing unit tests eliminates the need to run the project and navigate to the part you want to test with each change made. We will see throughout this chapter how beneficial unit tests will be to our project.

The other type of testing is called E2E testing. This type of testing requires the full app with its external resources, such as the API, to be available. The test ensures that a specific page or feature is behaving as expected in integration with all the software parts. E2E testing is important, but it's less common than unit tests; we will learn more about that in the *Introducing Playwright for E2E tests* section.

In this chapter, we will focus on unit testing, but before we get started practically, let's look at a summary of the benefits of writing great unit tests for your Blazor projects and components:

- Ensuring the reliability of the software by simulating scenarios that are not possible through traditional testing.

- Increasing the speed of development and discovering the code outcomes quicker, especially for complicated cases, and iterating quickly.

- Debugging becomes easier and more enjoyable because with good unit testing you can easily simulate what's going wrong at the unit level and debug it directly without needing to run the full app.

- Ensuring high-quality code, because writing unit tests forces the logic of the components to be written by following standards so that they can be tested.

- Ensuring that new changes do not break existing features. Since the testing is not manual but automated at the code level, and the unit tests can be run frequently, we can directly detect any new changes that cause existing features to fail by just running the existing tests.

Now, let's learn how to start writing tests for our **BooksStore** component.

Getting started with testing in Blazor with bUnit

Blazor doesn't have a native unit testing framework, so its great community came up with bUnit, which became the standard for testing Blazor apps.

bUnit allows you to write unit tests either in C# files or in Razor components, and it's compatible with all the common testing frameworks, such as **xUnit**, **NUnit**, and **MSTest**.

While testing a component, bUnit renders the target component in isolation and provides a full simulation, such as passing parameters, cascading values, and injecting services. It also simulates interactions with the component, such as clicking buttons or triggering event handlers.

The component being tested is known as the **Component Under Test** (**CUT**). The term CUT is what we will use throughout this chapter to name the component that we are testing. The term CUT is derived from the term **Service Under Test** (**SUT**), which is a known term in testing software overall, not just UI components.

After you simulate the component state you are testing, you can validate that the result is as expected. When it comes to the rendered UI of the component, bUnit allows you to match the desired HTML result with the rendering result of the component either as raw HTML or semantically. Comparing raw HTML means the desired output will match the rendered output if they are exactly the same text, just like comparing two strings. This will work fine in very simple cases, while in semantic comparison, the desired HTML will match the rendered HTML if they both have the same visualization when they are rendered inside a web browser. So, it's recommended to assert the results with semantic matching, as this allows flexible and advanced matching for validation and assertion, as we will see next.

> **Note**
>
> We will use xUnit as our testing framework in this chapter. xUnit is very popular due to its simple and clean syntax for writing tests. It supports data-driven tests by passing parameters and evaluates the output easily, and it has a rich set of extensions for assertions and result validations.

To get started with bUnit, we can either install the bUnit project template or create a testing project manually and reference the bUnit package inside. I prefer to install the bUnit template, as it has some default tests that show some of the package's capabilities out of the box:

1. Open the terminal and run the following command:

    ```
    dotnet new install bunit.template
    ```

2. Open the folder that has the Visual Studio solution file and the folders of projects inside the solution in the terminal.

3. Create a new bUnit project that uses xUnit as the testing framework and call it BooksStore. Tests:

    ```
    dotnet new bunit --framework xunit -o BooksStore.Tests
    ```

4. We need to add the BooksStore.Tests project to the BooksStore Visual Studio solution using the following command:

    ```
    dotnet sln BooksStore.sln add BooksStore.Tests
    ```

5. At the time of writing this book, the bUnit template creates a project that targets .NET 6.0; we are developing with .NET 7.0, so we need to upgrade. So, open the BooksStore.Tests. csproj file in Visual Studio and set the TargetFramework property to 7.0:

    ```
    ...
    <TargetFramework>net7.0</TargetFramework>
    ...
    ```

6. Lastly, add the BooksStore Blazor WebAssembly project as a reference to BooksStore.Tests so the test project can access the components and the code inside the BooksStore project:

    ```
    dotnet add BooksStore.Tests/BooksStore.Tests.csproj
        reference BooksStore/BooksStore.csproj
    ```

Our project is now ready to write unit tests. But before we start writing our own tests, the bUnit template contains a Counter.razor component and two test files, one written in a .cs file and the other written in a .razor file.

Let's open the CounterCSharpTests.cs file and see what's inside. The file has a class called CounterCSharpTest and it inherits from the TestContext class that's part of bUnit. TestContext provides all you need to create components under test, such as rendering, providing parameters, registering and injecting services, and interacting with the component elements via clicking a button or hovering the mouse over an element, and so on.

Inside the class, there are two methods, and each is decorated with the [Fact] attribute. [Fact] means this method is a unit test method that doesn't accept any parameters. When the test is executed, this method will run as it is, and it will either pass or fail without any prior conditions. Other attributes, such as Theory, can test the method with passing parameters. If you want to learn more about xUnit, please refer to their official site: https://xunit.net/.

Now, let's get into the methods inside CounterCSharpTests.cs:

```
. . .
[Fact]
public void CounterStartsAtZero()
{
    // Arrange
    var cut = RenderComponent<Counter>();

    // Assert that content of the paragraph shows counter
      at zero
    cut.Find("p").MarkupMatches("<p>Current count: 0</p>");
}
. . .
```

The preceding code is a unit test method called CounterStartsAtZero. As the name suggests, the Counter component should have a counter that starts from 0.

The first line renders the Counter component, then it validates the text inside the <p> of HTML and ensures it matches the text Current count: 0.

The second test method, ClickingButtonIncrementsCounter, validates the Counter component. The test will pass if the counter value increases after clicking the button:

```
[Fact]
public void ClickingButtonIncrementsCounter()
{
    // Arrange
    var cut = RenderComponent<Counter>();

    // Act - click button to increment counter
      cut.Find("button").Click();

    // Assert that the counter was incremented
    cut.Find("p").MarkupMatches("<p>Current count: 1</p>");
}
```

The second line finds the button inside the `Counter` component and uses the `Click` method to simulate clicking the button.

Finally, it validates that the current count will be `1`.

Those two methods demonstrate the big potential and capabilities of bUnit, including how it allows us to validate whether the component gets rendered as expected, or whether clicking a button inside the component results in the intended behavior and produces the expected results.

The other default file created with the bUnit template, called `CounterRazorTests.razor`, has the same tests but written in a Razor file to demonstrate how you can write tests in C# files or Razor.

Both ways are fine and will give you the same results, but usually, when writing tests for markup, it's better to host tests within a Razor file, as you have more flexibility while writing markup compared to writing within a string in a C# file. Additionally, it is easier to initialize and pass parameters to the CUT, whereas in C#, you need to write literal strings. In addition to using methods to pass parameters and initializing the CUT while in Razor, you can set up the CUT the same way you use the component and write markup in a normal Razor file, leveraging the power of the Razor editor in Visual Studio.

There are multiple ways to run unit tests in Visual Studio. For the one written in the C# file, you can either right-click on the test method and choose **Run Tests** or right-click on the class and click **Run Tests** to run all the tests inside the class.

To fully view, navigate, and run tests in Visual Studio written in C# or Razor, use the **Test Explorer** view. If you don't have it open by default, make sure to open it via the **View** menu and then choose **Test Explorer**. If you open **Test Explorer**, you should see the test as follows:

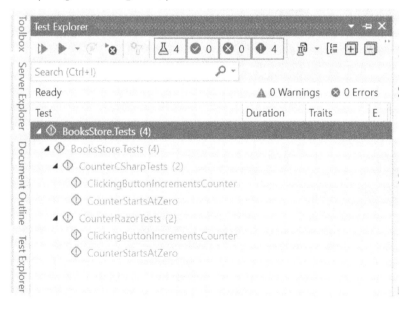

Figure 13.1 – Test Explorer in Visual Studio

To run all the tests in the view, click on the white and green arrow button in the top left. To run a specific test, select it from the view and click on the second green button with a single arrow. To make sure all is good, click on the first button (white and green), and then all the tests should have a green tick beside them.

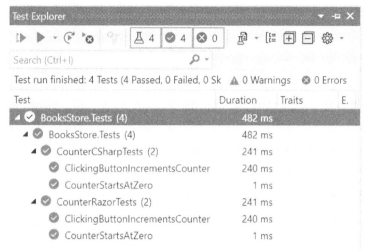

Figure 13.2 – Successfully run tests in Test Explorer

Now, the project is ready for us to start adding our own tests, which is what we will be doing next.

Writing component unit tests with bUnit

We have seen the basic tests written by default in the bUnit template. In this section, we will start creating our own tests. We'll write a test for the `ModalPopup` component we developed in the *Developing templated components* section in *Chapter 3, Developing Advanced Components in Blazor*.

Writing the first component unit test

`ModalPopup` is a good starting point for us as it doesn't have dependencies and it has some parameters, so we can learn how to pass parameters to a CUT.

Let's get started:

1. In the `BooksStore.Tests` project, add the `BooksStore` namespaces needed inside `_Imports.razor` so we don't need to reference them in each test file we have:

    ```
    . . .
    @using BooksStore.Shared
    @using BooksStore.Models
    @using BooksStore.Services
    ```

2. Create a new **Razor** component and name it `ModalPopupTests.razor`.

3. Remove any markup and make the component inherit from the `TestContext` class:

    ```
    @inherits TestContext

    @code {
    }
    ```

4. Inside the `@code` section, we will define the first test, which will test that the component renders the modal when we pass the `IsOpen` parameter the value `true`:

    ```
    . . .
    [Fact]
    public void ModalPopupWithIsOpenSet
      ToTrueShouldRendersCorrectly()
    {
        // Arrange
        var isOpen = true;
        var cut = Render(@<ModalPopup IsOpen="isOpen" />);
        // Assert
        var overlayDivs = cut.FindAll(".overlay");
        Assert.True(overlayDivs.Count > 0);
    }
    . . .
    ```

In the preceding method, first, we defined the `isOpen` variable and set it to true, and then we used the Render method and passed the `ModalPopup` component as a parameter. There is a Render method that renders a markup and returns an instance of type `IrenderedFragment`. `Render<TComponent>`. It also renders a fragment but returns `IrenderedComponent`. Both instances allow interaction with the rendered content, but `IRenderedComponent` allows for more control of the rendered component instance.

The other method to render a component is `RenderComponent<TComponent>`. You can pass the type of the component you want to render, and it will retrieve an instance of `IRenderedComponent`. Now, `RenderComponent<TComponent>` has an overload that accepts an array of `ComponentParameter` to help set the component parameters.

In summary, both `Render` and `RenderComponent` do the same job, but each gives a different level of control while creating the component and the type it returns. `Render` has a more flexible way of rendering the markup or the component, while `RenderComponent` is shorter, but passing parameters will require passing them as objects within an array.

To make sure that the component is rendered correctly, it should render a div with the class overlay, so we use the `FindAll` method and pass the `.overlay` element selector. `FindAll` will return all the elements that have the class overlay inside the rendered markup. So, if there is at least one div, that means the component is rendered successfully. `Assert.True` is the method that will assert the test result, as we passed the condition to it that the `overlayDivs` collection should have at least one item.

In **Test Explorer**, right-click on the new test and click on **Run**:

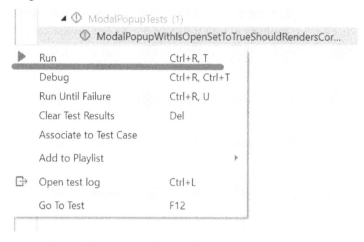

Figure 13.3 – Run the first ModalPopup component test

After running the test, if a green tick mark appears beside the name of the test, as shown in *Figure 13.2*, it means that the test should pass successfully!

Let's write one more test to validate that when the `IsOpen` parameter is set to `false`, no elements are rendered in the component. Again, in `ModalPopupTests.razor`, add the following test method:

```
...
[Fact]
public void
  ModalPopupWithIsOpenSetToFalseShouldRenderNothing()
{
    // Arrange
    var isOpen = false;
    var cut = Render(@<ModalPopup IsOpen="isOpen" />);

    // Assert
    Assert.True(cut.Nodes.Length == 0);
}
...
```

This test is similar to the previous one, but this validates the closing state of the modal where no elements should be rendered.

Testing components with RenderFragment

Let's look at an example where we can learn how to match partial markup within the rendered component. We know that ModalPopup receives ChildContent and FooterContent render fragments that are rendered inside the modal in the open state, so we need to make sure that ModalPopup is rendering ChildContent and FooterContent correctly. In the next test, we will provide the ModalPopup component with the ChildContent parameter and make sure it's rendered as expected:

```
[Fact]
public void
  ModalPopupWithChildContentShouldRenderCorrectly()
{
    // Arrange
    var cut = Render(
    @<ModalPopup IsOpen="true">
        <ChildContent>
            <h2>Testing ModalPopup</h2>
        </ChildContent>
    </ModalPopup>
    );

    // Assert
    cut
    .Find(".modal-body")
    .MarkupMatches(@<div class="modal-body"><h2>Testing
        ModalPopup</h2></div>);
}
```

The preceding test passes ChildContent and sets IsOpen to true, so based on the component logic, the ChildContent render fragment should be rendered inside a div with the modal-body class. The MarkupMatches method tries to match the expected rendered result of the div element with the class modal-body. The matching of MarkupMatches happens semantically, which means the expected string and the rendered markup shouldn't necessarily be exactly equal, but they should both have the same output if they are rendered within the browser. For example, let's assume the rendered component has the following markup:

```
<div class="card shadow-1 corner-radius"><h1>Title
    here</h1></div>
```

Using `MarkupMatches`, we can compare the preceding markup with the following markup, and they will be equal, despite having spaces or class order differences. Both will produce the same output in the browser:

```
<div        class="card corner-radius shadow-1">
<h1>Title here</h1>
</div>
```

Testing components with interaction

The last test we will write in this section is a test that involves an interaction, such as clicking on a button, updating the parameters during the test execution, and triggering re-rendering for the component. If `ModalPopup` is open, it has an **X** button that closes it. The upcoming test will render `ModalPopup` in the closing state, then set `IsOpen` to `true`, simulate clicking the **X** button, and finally, ensure that the modal is closed and no elements are rendered:

```
...
[Fact]
public void ClickXButtonShouldCloseTheModal()
{
    // Arrange
    var cut = RenderComponent<ModalPopup>();

    // Act
    cut.SetParametersAndRender(parameters =>
    {
        parameters.Add(p => p.IsOpen, true);
    });
    cut.Find(".close-button").Click();

    // Arrange
    Assert.True(cut.Nodes.Length == 0);
}
...
```

This time, we used the `RenderComponent` method instead of `Render` and it set `ModalPopup` as the component type. The `SetParametersAndRender` method allows us to update the component parameter values and trigger a re-rendering for the component. After `IsOpen` is set to `true`, the modal should be rendered and a `<span>` tag with the `close-button` class should be found inside the markup, so we used the `Find` method and then the `Click` method, which simulates clicking that element. Lastly, during the assertion, we validate that there are no nodes anymore in the rendered component after clicking the button, as the modal should be closed.

We've created four tests that cover various scenarios including two tests to validate the initial render state of the `ModalPopup` component, another one that includes passing parameters, and one to validate the component logic by interacting with it. This should give you a strong foundation to write additional tests that cover a range of cases. The `ModalPopup` component is straightforward and doesn't require any injected services, and its logic remains unaffected by the authentication state. However, for more advanced scenarios, you'll need to explore more complex topics, which we'll cover in the next section.

Mocking and faking tests in Blazor and bUnit

Writing stable unit tests requires isolation for the unit under test as the unit tests should run frequently in different environments and on different machines. So, to keep the tests reliable, they should be written in isolation from external factors such as communicating with an API or asking for an access token.

Mocking and faking are the keys to simulating the behavior of the dependencies of the unit under test. For example, the **Index** page depends on `IBooksService`, and the implementation we have is called `BooksHttpClientService`, which sends requests to the API to fetch the books needed for the **Index** page. If we want to test `IBooksService`, we need to mock it and write a fake implementation for the `GetAllBooksAsync` method so the unit test can rely on this fake service instead of the actual API.

In order to mock services in .NET, there are many available packages, but **Moq** is the most common one, and it's the one we will be using in this section.

The mocked and fake services and implementations are known as **test doubles** in the world of software testing, and bUnit has a set of test doubles for the most common Blazor services, such as `NavigationManager`, fake authentication and authorization, and more.

Writing tests with Moq

A good example to start with is the `BookDetails` component. This component represents a page that renders the book's details. The component depends on `IBooksService` to retrieve a book by ID.

So, let's see how this works:

1. In order to test the `BookDetails` component, we need to install the `Moq` package before writing the tests.

    To install the `Moq` package, either install it directly from NuGet in Visual Studio or run the following .NET CLI command within the folder of the `BooksStore.Tests` project:

    ```
    dotnet add package Moq
    ```

2. In `_Imports.razor`, add the Moq namespace:

    ```
    @using Moq;
    ```

3. In the tests project, create a new Razor component called `BookDetailsTests.razor`.

4. Add a namespace reference for `BooksStore.Pages` and let the component inherit from the `TestContext` class:

```
@using BooksStore.Pages.UserPages
@inherits TestContext

@code {

}
```

5. The test we will write should test the perfect scenario of the component that simulates the `Book` object being retrieved correctly and the component rendering the data correctly. The `BookDetails` component receives a `BookId` parameter in the form of a string, and then the component calls the `GetBookByIdAsync` method inside `IBooksService` to retrieve the book's details. For this test, we need to mock `IBooksService` and register the mocked instance in the DI of `TestContext`:

```
[Fact]
public void
  BookDetailsWithValidBookIdShouldRenderCorrectly()
{
    // Arrange
    var mockedBookService = new
      Mock<IBooksService>();
mockedBookService
    .Setup(x =>
      x.GetBookByIdAsync(It.IsAny<string>()))
.Returns(Task.FromResult<Book?>(new Book
    {
        Id = Guid.NewGuid().ToString(),
        Title = "Test Book",
        AuthorName = "Test Author",
        Description = "Test Description",
        Price = 10
    }));

            Services.AddScoped<IBooksService>(sp =>
              mockedBookService.Object);
}
```

The preceding code mocks the `IBooksService`, then it sets up the `GetBookByIdAsync` method using the `Setup` method. `It.IsAny<string>()`, which is passed as a parameter, is a method provided by the Moq framework. It's used when the parameter value is not important for the test, so as the name indicates, it's any string. After that, the `Returns` method is used to define the value to return for the mocked `GetBookByIdAsync`. Finally, we register `IBooksService` in the DI container of `TestContext` using the `Services` property provided by bUnit.

6. Now, we can render the `BookDetails` component and validate the result:

    ```
    ...
    var cut = Render(@<BookDetails BookId="1" />);

    // Assert
    cut.Find("h2").MarkupMatches(@<h2>Test Book</h2>);
    ```

    That's it! Now, when you run the test, `IBooksService` will be injected and `GetBookByIdAsync` will be called, and the mocked book we created in *step 5* will be returned. Then, the component will be rendered successfully, and to ensure it's rendered correctly, we check the `<h2>` tag that prints the book's title, which is `Test Book` in this case.

Next, we will look at another kind of component testing, which is testing the components that require authentication. Let's see how bUnit provides an easy way to fake authentication and authorization in the test context.

Faking authentication and authorization in bUnit

In the *Implementing authorization and advanced authentication features* section in *Chapter 9, Authenticating and Authorizing Users in Blazor WebAssembly*, we modified the `BookCard` component to show the **Add to Cart** button only if the user is authenticated and located in the UK, based on a custom policy called `UK_Customer`.

When it comes to testing scenarios involving authentication and authorization, bUnit makes the process much easier than you might expect. That's because it comes with a built-in fake authentication and authorization infrastructure that you can readily use. With the bUnit APIs, you'll have full control over faking the authentication or authorization process, with the ability to set claims, roles, policies, and more, in a very straightforward manner. This can greatly simplify your testing efforts, enabling you to focus on verifying your components' behavior in different security scenarios without having to worry about the complexities of setting up real authentication and authorization infrastructure.

Let's write two unit tests for the `BookCard` component. The first will render the card without authentication, so there should be no **Add to Cart** button and it should render a message saying that only customers within the UK can buy the book. The second is for when the user satisfies the `UK_Customer` policy, so the **Add to Cart** button should appear:

1. Create a new Razor component in the `BooksStore.Tests` project, name it `BookCardTests. razor`, and make it inherit from `TestContext`:

    ```
    @inherits TestContext
    ```

```
@code {

}
```

2. Create a Book instance variable called _book to be passed for the BookCard component tests:

```
@code {

    private Book _book = new Book()
    {
        Id = Guid.NewGuid().ToString(),
        AuthorName = "Test Author",
        Description = "Test Description",
        Price = 10,
        Title = "Test Book"
    };
}
```

3. Create a test method for the UK customer that validates whether the Add To Cart button should be rendered:

```
...
[Fact]
public void
  UsersWithUK_CustomerPolicyShouldSeeAddtoCartButton()
{
    // Arrange
    var authContext = this.AddTestAuthorization();
    authContext.SetPolicies("UK_Customer");
    var cut = Render(@<BookCard Book="_book" />);

    // Assert
    cut.Find("button").MarkupMathces(@<button
      class="main-button">Add to Cart</button>);
}
...
```

The AddTestAuthorization extension method is the method that's responsible for the fake authorization in bUnit. Using the authContext variable, we can call the SetPolicies method to set a policy (UK_Customer for our case), SetClaims to set fake claims, SetRoles to set roles, and more. You can learn more about faking authorization in bUnit at https://bunit.dev/docs/test-doubles/faking-auth.html.

4. Write another test method without authorization, as we should expect the same component to not render the Add to Cart button and render a `<p>` tag with a message inside, as follows:

```
...
[Fact]
public void UsersWithoutUK_
  CustomersPolicyShouldNotSeeAddToCartButton()
{
    // Arrange
    this.AddTestAuthorization();
    var cut = Render(@<BookCard Book="_book" />);
    var buttons = cut.FindAll("button");

    // Assert
    Assert.Empty(buttons);
    var textContent = cut.Find("p.text-
      center").TextContent;
    Assert.Equal("Books only available for sale in UK
      for the time being", textContent);
}
...
```

In the second test, we didn't provide the UK_Customer policy, so to assert the result of the test, we used FindAll to find all the buttons in the markup and assert that the list is empty, and we validated that the markup of the `<p>` tag with the text-center class is rendered and shows the required text. Here, we used the TextContent property of the node. TextContent has the value of the content inside the tag. Then, we made sure that the content of the `<p>` tag is equal to the expected text using the Assert.Equal method.

That's fun, right? Nothing in developing software is more enjoyable than writing a lot of tests and having them pass every time you run them. During the last few pages, we have written 11 tests. That's so satisfying and, using what we have learned in this chapter so far, you can write more tests for the rest of the pages and components. The following screenshot shows the state of **Test Explorer** at the end of this section:

Figure 13.4 – Test Explorer with all the tests we've written

bUnit is a very powerful framework. We have seen so far how it helps with writing efficient tests and how much flexibility it provides with writing and asserting unit tests for Blazor components. What we have seen so far is not everything; bUnit also provides much more in the shape of features, capabilities, mocks, and so on, and to discover them all, check out their simple and easy-to-follow documentation: `https://bunit.dev/docs/getting-started/index.html`.

> **Note**
>
> Component unit testing is preferable to test the flow and observe the state of the component during its lifetime and while interacting with it. It's not recommended to test a specific parameter, private field values, or specific methods. Testing has to be done for the overall flow and the purpose of the component in general.

We have made good progress in writing unit tests. In the next section, we will briefly introduce the **Playwright** testing framework to write E2E tests for Blazor WebAssembly.

Introducing Playwright for E2E tests

In E2E tests, we need to ensure that our system is fully behaving as expected, including the interaction with external services, such as the web API.

Even though unit testing is more crucial than E2E testing, a great testing strategy includes all different types of tests, including unit tests, integration tests, E2E tests, and performance tests. The reason why automated E2E tests are less important than unit tests is that all the components have to be available and running. In our case, every time you want to run E2E tests, the API has to be up and running, and a sandbox environment has to be available, otherwise the tests will keep adding, editing, or deleting data if the API doesn't support test data. E2E tests are also slower to run, so running them all the time, as we can with unit tests, is not preferable.

E2E tests complement rather than replace unit tests for the product's overall testing. E2E testing is not meant to replace unit testing. Instead, it makes sure that all the features and app infrastructure are working together as expected. This includes the services in the DI container, the API calls, the serialization and deserialization of the data, and so on.

To write E2E tests, Playwright is the ideal choice for Blazor WebAssembly. Playwright for .NET is an open source library that's currently maintained by Microsoft. It provides APIs that automate browser tasks, which allows developers to write browser automation tests in .NET. Playwright supports most modern browsers, such as Chromium and Firefox, in addition to supporting mobile emulations. With Playwright, the .NET tests can interact with the site inside the browser, such as navigating between links and filing and submitting forms.

An example test in Playwright can be built with the following steps:

1. Initialize the Playwright instance.
2. Define the browser to use.
3. Navigate to a page inside the browser.
4. Click on a button on the page.
5. Ensure that the clicked button achieved the desired goal, such as navigating to another page or submitting a request successfully.

From the test context provided, you can see that the full app is included here, unlike in unit tests, where the CUT is the only part we focus on.

Playwright currently supports *NUnit* and *MSTest* testing frameworks and, luckily, it has great documentation that explains in detail how to get started and how to utilize each feature provided by the package. If you are interested in exploring Playwright for .NET, feel free to check out their site: `https://playwright.dev/dotnet/docs/intro`.

This should give you a quick introduction to E2E tests and what to use to start writing them.

Writing tests is not only about knowledge and experience; consistency and discipline play big roles here. Taking tests seriously from the early stages of the project helps you write high-quality components faster and ensures the reliability of the code through the process of development.

Summary

Throughout this chapter, we have learned about testing software and especially Blazor WebAssembly components. First, we introduced the concept of testing, why it's important, and the difference between unit testing and E2E testing. Then we introduced bUnit and set up the testing project. After that, we discovered the default tests written in a bUnit testing project template before starting to write our own, while learning gradually about bUnit's features. While writing tests, we saw the need for mocking and faking services and features to satisfy the CUT. We learned about the Moq framework and the built-in mocking and faking services within bUnit. Finally, we introduced the Playwright library for .NET, which we can use to write E2E tests in Blazor WebAssembly.

After completing this chapter, you should be able to do the following:

- Understand the importance of testing software

- Be familiar with the bUnit library and its features

- Write unit tests for different kinds of components in different scenarios

- Mock and fake services when needed using Moq and bUnit

- Understand the need for E2E tests and how to get started writing them

You have developed your app and written a lot of tests for it that pass. The client, friends, family members, or the users who know what you are working on are excited to get their hands on what you are developing. So, it's time to send them a URL and ask for their feedback. Let's push the project and make it accessible online!

Further reading

- Testing Razor components in ASP.NET Core Blazor: `https://learn.microsoft.com/en-us/aspnet/core/blazor/test?view=aspnetcore-7.0`

- bUnit documentation: `https://bunit.dev/docs/getting-started/index.html`

- Playwright for .NET: `https://playwright.dev/dotnet/docs/intro`

- Testing Blazor applications with Playwright: `https://learn.microsoft.com/en-us/events/dotnetconf-2022/testing-blazor-applications-with-playwright`

14

Publishing Blazor WebAssembly Apps

After working so hard on your app and after going through all the difficult moments while debugging an error or learning a new technique to build a new feature, the feeling of seeing your app online is priceless, regardless of how big or simple it is.

Finally, it is that time for us. It's time to put the app we have built so far online and make it accessible via a public URL and have a party after that.

In this chapter, we will introduce some final checks to do for your Blazor WebAssembly app before the release; then, I will introduce you to **Blazor WebAssembly ASP.NET Core Hosted**. We will also cover the concepts of **continuous integration** (**CI**) and **continuous delivery** (**CD**). Finally, we will push the project to **Azure App Service** and **Azure Static Web Apps** using the **Azure portal** and **GitHub Actions**.

In this chapter, we will go over the following topics:

- Blazor WebAssembly app prerelease checks
- Getting to know the Blazor WebAssembly ASP.NET Core Hosted app
- Publishing the project to Azure App Service
- Publishing the project to Azure Static Web Apps

Technical requirements

The Blazor WebAssembly code and the API project used throughout this chapter are available in the book's GitHub repository in the *Chapter 14* folder:

```
https://github.com/PacktPublishing/Mastering-Blazor-WebAssembly/
tree/main/Chapter_14/
```

Also, the full version of the Blazor WebAssembly ASP.NET Core `BooksStore` project can be found in the following folder within the GitHub repository:

```
https://github.com/PacktPublishing/Mastering-Blazor-WebAssembly/
tree/main/Complete/AspNetCoreHosted
```

An Azure account is required to use Azure App Service and Azure Static Web Apps, to which we will publish the Blazor project. If you don't have an Azure account already, you can create a free account via the following link: `https://azure.microsoft.com/en-us/free/`.

Blazor WebAssembly prerelease final checks

Before releasing the Blazor WebAssembly app, it's useful to know how the process happens and what things we need to configure. We will look at **Ahead Of Time** (**AOT**) compilation, trimming, URL rewriting, and Blazor time zone support to give you a better idea of whether there is anything you need to configure related to those topics. Let's get started.

AOT compilation

By default, Blazor WebAssembly apps run in the browser via a .NET **Intermediate Language** (**IL**) interpreter implemented in WebAssembly.

> **Note**
>
> IL is the binary result of the compiled high-level .NET code, and you can learn more about IL at the following link: `https://learn.microsoft.com/en-us/dotnet/standard/managed-code#intermediate-language--execution`.

Since .NET code is interpreted, Blazor WebAssembly apps can be slightly slower than .NET apps that run on the server, particularly when performing complex calculations or manipulating large amounts of data. However, enabling AOT compilation during publishing can mitigate this performance issue by compiling the .NET code to WebAssembly before the runtime in the browser, eliminating the need for interpretation. This feature is called *Ahead of Time* because it compiles the .NET code to WebAssembly ahead of runtime in the browser. While AOT can improve performance, it does come with a cost. Enabling AOT will increase the app's initial size, resulting in a larger publishing output – twice the size of publishing without AOT. This can increase the time it takes to load the app for the first time. So, it's important to consider using AOT only when your app is doing heavy client-side computations – for example, a diagram editor app, where the app is mostly doing SVG manipulations on the client side. For apps that mostly depend on API calls and render the data in the client, as in the `BooksStore` project, enabling AOT will be a huge downside as it won't add any performance upgrade for the client and it will increase the app's initial download size.

To enable AOT, open the `.csproj` file of your Blazor project and add the following property:

```
<PropertyGroup>
  <RunAOTCompilation>true</RunAOTCompilation>
</PropertyGroup>
```

That's all you need to enable AOT. Later, when we publish the app or build it for the release, the app will be compiled into WebAssembly. AOT compilation isn't used during the development process because the process of compilation takes a couple of minutes to finish. It doesn't make sense to wait that long every time you run the project.

Trimming

Trimming is another task that's performed during the publishing phase and it happens automatically. Trimming basically removes unused code from the produced IL code, which reduces the size of the app significantly.

If your app is using reflection features in C#, the trimming operation cannot determine the actual types you are targeting, so they may be trimmed, which could result in uncertain behavior. For this specific scenario, you might need to configure the trimmer to ignore trimming a specific assembly or even the level of trimming in general.

To ignore trimming specific assembly code, you can add the following to your project file:

```
<ItemGroup>
  <TrimmerRootAssembly Include="Name_Of_Your_Assembly" />
</ItemGroup>
```

`RootAssembly` means all the assembly and its dependencies won't be trimmed.

Or you can fully disable the trimming operation using the `<PublishTrimmed>` property:

```
<PublishTrimmed>false</PublishTrimmed>
```

You can learn more about how to control the trimming operation at the following link: `https://learn.microsoft.com/en-us/dotnet/core/deploying/trimming/trimming-options?pivots=dotnet-7-0`.

> **Note**
>
> When AOT is enabled, the output won't be trimmed as the result is native WebAssembly code, and that's another reason why the AOT-compiled app is large.

Compression

When publishing a Blazor WebAssembly app, the output is compressed to reduce the app size and increase the startup time of the app. Also, compression happens by default.

To disable compression, which is undesirable (as disabling it will increase the app size significantly), you can add the `BlazorEnableCompression` property to the project file:

```
<PropertyGroup>
<BlazorEnableCompression>false</BlazorEnableCompression>
</PropertyGroup>
```

The only scenario in which you may need to disable compression is when you publish your app to a host that already serves compressed files by itself, such as the **AWS CloudFront Content Delivery Network** (**CDN**). In such scenarios, also having the app files compressed by Visual Studio will increase the overhead for the WebAssembly runtime. You can learn more about how to publish a Blazor WebAssembly project to AWS and use the AWS CloudFront CDN to serve your files here: `https://aws.amazon.com/blogs/developer/run-blazor-based-net-web-applications-on-aws-serverless/`.

URL rewriting

When you publish a standalone Blazor WebAssembly project, there will be a single HTML file called `index.html`. This file is downloaded to the browser and then bootstraps the app. For example, if your app is accessible through the URL `https://bookstore.example`, when the user navigates to that URL, the server will serve the `index.html` file by default, which is good.

While the user is navigating through the pages on the client side, everything should be fine. For example, if the user clicks on the **Add Book** link in **NavMenu**, the user will be redirected to `https://bookstore.example/book/form` correctly as the routing and navigation process is happening on the client side within the browser.

However, a problem may occur if the user navigates to `https://bookstore.example/book/form` directly from the browser's address bar. In this case, the browser will send the request to the app host, and there is nothing mapped to `/book/form`, so the result will be a `404` page.

The solution for this issue is to configure the host to map any request to `index.html`.

We will see how to configure the fallback URL to `index.html` when we deploy the project to Azure Static Web Apps.

If you are planning to host your Blazor WebAssembly project on **IIS**, check out the following link: `https://learn.microsoft.com/en-us/aspnet/core/blazor/host-and-deploy/webassembly?view=aspnetcore-7.0#iis`.

Disabling time zone support

By default, the Blazor WebAssembly runtime includes a data file to make the time zone data correct. Disabling this feature will decrease the initial app size, in addition to improving the performance of some operations such as parsing and formatting dates.

To disable the time zone feature in Blazor, you can add the following property and set it to `false` in the `.csproj` file:

```
<PropertyGroup>
   ...
<BlazorEnableTimeZoneSupport>false</BlazorEnableTimeZoneSupport>
   ...
</PropertyGroup>
```

Keep in mind that disabling time zone support will make the app use the browser's time zone information.

In this section, we talked about the publishing operation and what you can configure during it.

In the next section, we will look at another way to build and host a Blazor WebAssembly project, called Blazor WebAssembly ASP.NET Core Hosted. Let's get into it.

Introducing Blazor WebAssembly ASP.NET Core Hosted

Throughout this book, we fully focused on the Blazor WebAssembly project; even when we introduced the API, I kept it away so you could just focus on what matters in Blazor WebAssembly within the browser.

What we have created during this book is called a Blazor WebAssembly standalone project, which means a project that serves itself and will be published without any server or backend to serve it.

.NET already offers an out-of-the-box template for a Blazor WebAssembly project called **Blazor WebAssembly ASP.NET Core Hosted**. This template creates a Blazor WebAssembly project, a shared class library project, and an ASP.NET Core API directly inside the solution. The most interesting part of this type of project is that the API project is responsible for serving the Blazor WebAssembly app to the client. In other words, you only need to host the server project and the Blazor WebAssembly project will be referenced inside it.

The Blazor WebAssembly ASP.NET Core Hosted template and hosting model are particularly useful when you're building a project from scratch and need to develop both the API and the Blazor app together as a single unit. This approach is different from our current project, where the API and Blazor WebAssembly are hosted separately, resulting in two different URLs. Another big feature of using the ASP.NET Core Hosted model is the server-side prerendering. Blazor WebAssembly renders the UI and the components on the client side in the browser, but using server-side prerendering, the initial HTML content of the app is rendered on the server and sent back to the client in the first response. This improves the overall user experience and performance, in addition to giving you the ability to utilize **Search Engine Optimization** (**SEO**) in your Blazor WebAssembly apps.

> **Note**
>
> Blazor WebAssembly server-side prerendering can be beneficial in many scenarios including a blog app, e-commerce, or even our `BooksStore` project. Prerendering allows you to add SEO to your app, and the app pages will load faster just like a traditional web app, which is essential for such scenarios.
>
> To not be out of topic for publishing and hosting a Blazor app, I added a prerendering sample for the GitHub repository of the book, which can be found at the following link: `https://github.com/PacktPublishing/Mastering-Blazor-WebAssembly/tree/main/Chapter_14/AspNetCoreHostedProject/PrerenderingSample`. Additionally, Microsoft has a great documentation page that explains the process step by step here: `https://learn.microsoft.com/en-us/aspnet/core/blazor/components/prerendering-and-integration?view=aspnetcore-7.0&pivots=webassembly`.

To create a new Blazor WebAssembly ASP.NET Core Hosted, we need the following:

- *Visual Studio*: After choosing the Blazor WebAssembly project template, make sure to check **ASP.NET Core Hosted**.

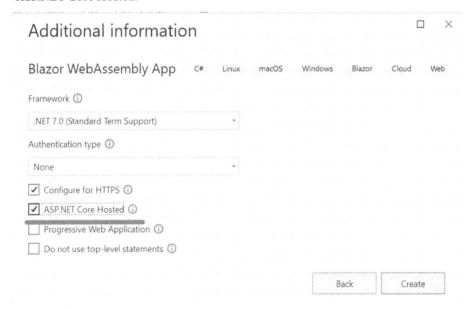

Figure 14.1 – Creating a Blazor WebAssembly ASP.NET Core Hosted project

- *.NET CLI*: Use the following command:

```
dotnet new blazorwasm --hosted -o <project-name>
```

The project has the following structure in Visual Studio:

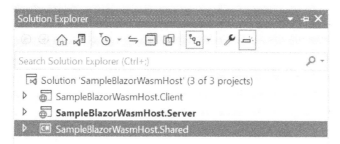

Figure 14.2 – Blazor WebAssembly ASP.NET Core Hosted project structure

For the `BooksStore` project, in the GitHub repository of this book, you can find another version of the current project alongside the API that is Blazor WebAssembly ASP.NET Core Hosted.

> **Note**
>
> The choice of project type comes back to the API. If you are developing the API, it's better to go for Blazor WebAssembly ASP.NET Core Hosted as you need to host one app, and the project template accelerates the process for you. If your project doesn't require an API or requires third-party APIs or an existing hosted API, then you need to proceed with the standalone hosting model.

The reason why I provided two versions of the project (one as a standalone project and another as an ASP.NET Core Hosted project) was to show you how to publish both types of projects. So, next, we will deploy the ASP.NET Core Hosted version to Azure App Service and have a URL to access our app through.

Publishing Blazor WebAssembly to Azure App Service

In this section, we will publish the Blazor WebAssembly ASP.NET Core Hosted version to the cloud, specifically Azure, and the hosting service will be **Azure App Service**.

Azure App Service is a **Platform-as-a-Service (PaaS)** solution that allows the deployment of web apps at scale. The service is fully managed by Azure and it supports different programming languages and frameworks, in addition to supporting containers.

Azure App Service provides multiple deployment options to push your app to it:

- **Continuous deployment**: Azure App Service supports continuous deployment from multiple sources, such as *GitHub*, *Azure DevOps*, and *Bitbucket*. This means that whenever changes are pushed to the source code repository, the application will be built and deployed automatically to Azure App Service. You can learn more about continuous deployment for Azure App Service at `https://learn.microsoft.com/en-us/azure/app-service/deploy-continuous-deployment?tabs=github`.

- **FTP deployment**: FTPS deployment is also supported by Azure App Service. You can upload the project using an FTP or FTPS client. Learn more at `https://learn.microsoft.com/en-us/azure/app-service/deploy-ftp?tabs=portal`.

- **Container deployment**: Azure App Service fully supports the deployment of containerized applications by providing built-in support for **Docker**. To learn more, check out `https://learn.microsoft.com/en-us/azure/app-service/deploy-ci-cd-custom-container`.

- **Web Deployment via Visual Studio**: With this method, you directly publish your project to Azure App Service from within Visual Studio.

In this exercise, we will use the Visual Studio deployment option. From within Visual Studio, you can create a new Azure App Service instance and upload directly to it, but we will create the required resources on Azure and then just use Visual Studio to publish.

Make sure to have the `BooksStore` Blazor WebAssembly ASP.NET Core Hosted version cloned on your machine to be able to proceed with this exercise. You can find the project in the following folder: `https://github.com/PacktPublishing/Mastering-Blazor-WebAssembly/tree/main/Complete/AspNetCoreHosted`.

If you are following this book with your own project idea and a web API is not needed, you can still use the steps here to deploy the Blazor WebAssembly app to Azure App Service.

Let's get started with publishing the project:

1. Open the browser and navigate to the Azure portal via `https://portal.azure.com`.

2. On the home page, open the left menu and click on **Create a resource**:

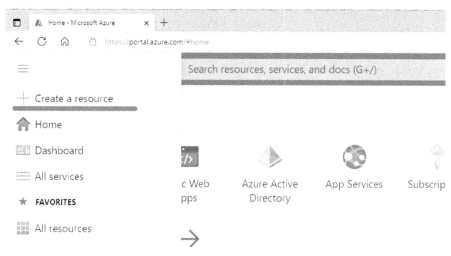

Figure 14.3 – Create a resource in the Azure portal

3. On the **Create a resource** page, choose the **Web App** service:

Figure 14.4 – The Web App Azure service

4. On the **Create Web App** page, choose your subscription, create a new resource group with a unique name, define the name of your Azure App Service (it must be unique globally), choose **Code** as the **Publish** method, select **.NET 7** for **Runtime stack**, select **Windows** for **Operating System**, choose an Azure region where the app will be hosted, and finally, for **Pricing plan**, make sure to choose **Free F1 (Shared Infrastructure)**. The following screenshot shows the values I have used:

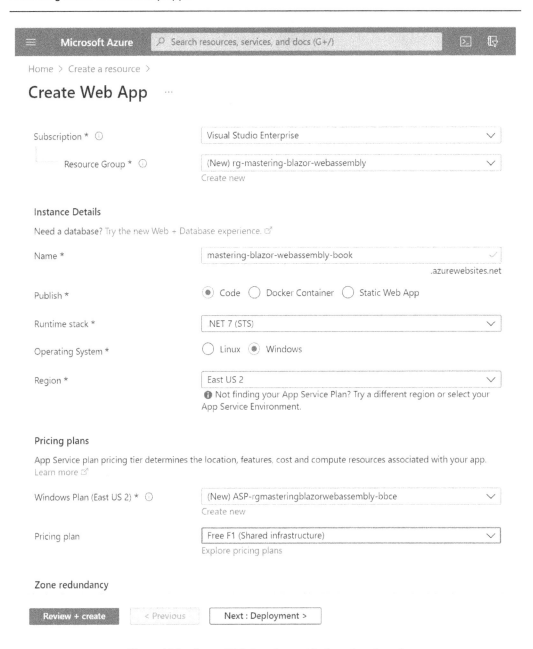

Figure 14.5 – Create Web App form with the values I used

5. Click on **Review + create**, then click **Create** again after the validation finishes.

6. Wait until the Azure App Service resource is created. Once Azure finishes the deployment, you should see the following UI:

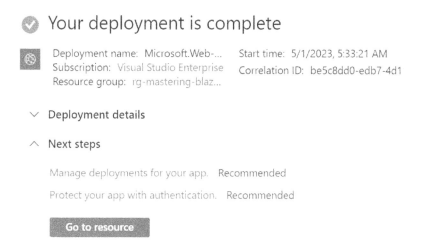

Figure 14.6 – Azure App Service is deployed

7. Click on **Go to resource** to open the **Azure App Service** resource page of our project, and then click on **Download publish profile**:

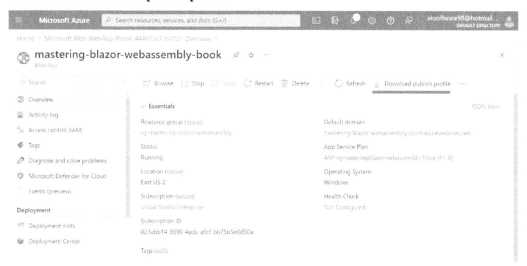

Figure 14.7 – Download publish profile button

Now, we can open the solution in Visual Studio and start deploying the project to the created Azure App Service.

8. Because the project is ASP.NET Core Hosted, we need only to deploy the server project, so right-click on the **BooksStore.Server** project and click on **Publish…**:

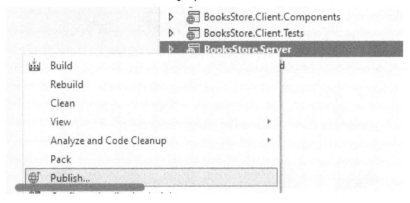

Figure 14.8 – The Publish… button for the server project

9. In the **Publish** dialog, choose **Import Profile** and browse the downloaded profile from *step 7*.

10. Now, the project is ready to be published. Just click the **Publish** button and Visual Studio will build the project in the **Release** mode and start the deployment.

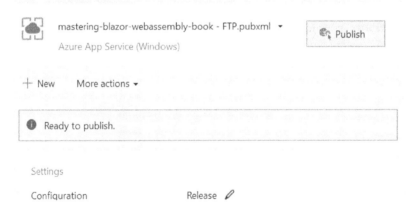

Figure 14.9 – Publish to Azure App Service profile

Done! Once the app is built and published, the browser will automatically open and navigate to your new app. The URL of the app will be https://{AZURE_APP_SERVICE_NAME}.azurewebsites.net, so in our example, the URL of the app is https://mastering-blazor-webassembly-book.azurewebsites.net.

Congrats! You have published your first Blazor WebAssembly project. The feeling is amazing, but we are not done yet. We will discuss another interesting service called Azure Static Web Apps for publishing our Blazor WebAssembly standalone project.

Publishing Blazor WebAssembly to Azure Static Web Apps

Azure Static Web Apps is a recently introduced Azure resource that simplifies the building and deployment of static web apps. However, it's important to note that it's not suitable for publishing a Blazor WebAssembly ASP.NET Core Hosted app as it doesn't provide a server that supports .NET. Instead, it's designed to serve static web files and projects and is optimized for this type of application. One of its key benefits is its ability to distribute static files globally, enabling users to access an app quickly from the nearest geographic location.

One of the great features of Azure Static Web Apps is its automatic CI and CD pipeline when using GitHub or Azure DevOps. This means that whenever you submit any code changes to the repository, the pipeline is triggered to build and deploy the updated code.

Make sure that you have the latest version of the `BooksStore` project; you can clone it from the following URL:

```
https://github.com/PacktPublishing/Mastering-Blazor-WebAssembly/
tree/main/Chapter_14/Final/
```

Now, let's deploy our Blazor WebAssembly project to a new instance of Azure Static Web Apps:

1. Before we start publishing, we need to set the right URL for the API because, now that we are publishing the Blazor WebAssembly alone, we need to set the API URL used in `HttpClient` to the correct API. You can set the API link within the `appsettings.json` file inside the `wwwroot` folder of the Blazor project. To get the API link, you can either deploy the API project yourself the same way we deployed the Blazor project in the previous step or you can just set it to `https://mastering-blazor-webassembly-book.azurewebsites.net`; I have published the API at this URL, so you can just use it as it is:

    ```
    {
        "ApiUrl": "https://mastering-blazor-webassembly-
          book.azurewebsites.net"
    }
    ```

2. Make sure to create a new GitHub repository, clone it, copy the project files to it, then push it, so the source code is available on GitHub. Or just navigate to the following GitHub repository and click on **Fork**: `https://github.com/aksoftware98/mastering-blazor-webassembly-project`.

3. Navigate to the Azure portal via `https://portal.azure.com`.

4. From the left menu, click on **Create a new resource**.

5. Search for `azure static web apps`, and then choose the **Static Web App** service:

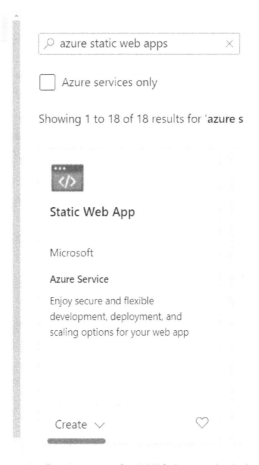

Figure 14.10 – Creating a new Static Web App service in Azure

6. On the **Create Static Web App** page, choose an existing resource or create a new one. Then, set the name of the resource, the **Free** plan type, and the region for the Azure Functions API, which we don't have for our project, but the Static Web App asks for it nonetheless:

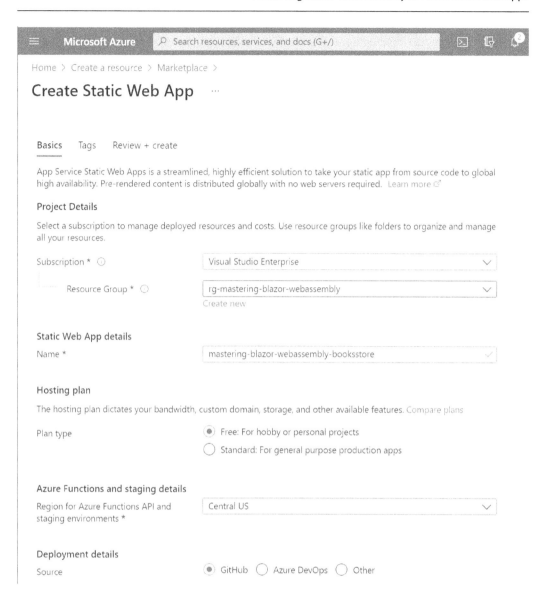

Figure 14.11 – Create Static Web App details

7. In **Deployment details**, choose **GitHub**, then click on **Sign in with GitHub** if you are asked to.

8. For the GitHub details, choose your organization, then the **Repository** name, and the **Branch** name, as shown here for my GitHub account:

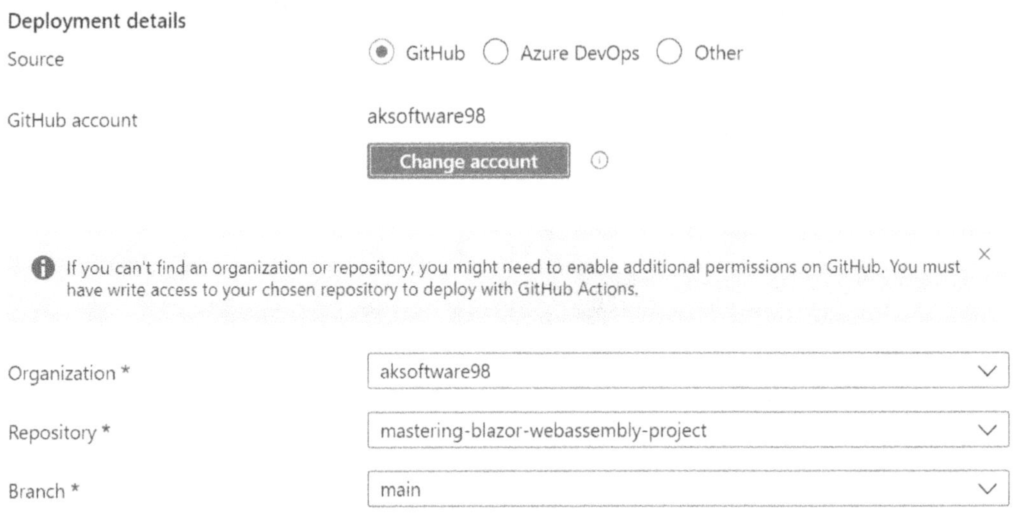

Deployment details

Source   ◉ GitHub   ◯ Azure DevOps   ◯ Other

GitHub account   aksoftware98

[**Change account**]   ⓘ

ⓘ If you can't find an organization or repository, you might need to enable additional permissions on GitHub. You must   ✕
have write access to your chosen repository to deploy with GitHub Actions.

Organization * | aksoftware98   ⌄ |

Repository * | mastering-blazor-webassembly-project   ⌄ |

Branch * | main   ⌄ |

Figure 14.12 – Deployment details of the Static Web App resource

9. The last part we need to fill out is **Build Details**. Choose **Blazor** for **Build Presets**, then set **App location** to /src/BooksStore because this is the folder that contains the BooksStore Blazor project file. Leave the **Api location** field empty, and leave **Output location** as it is:

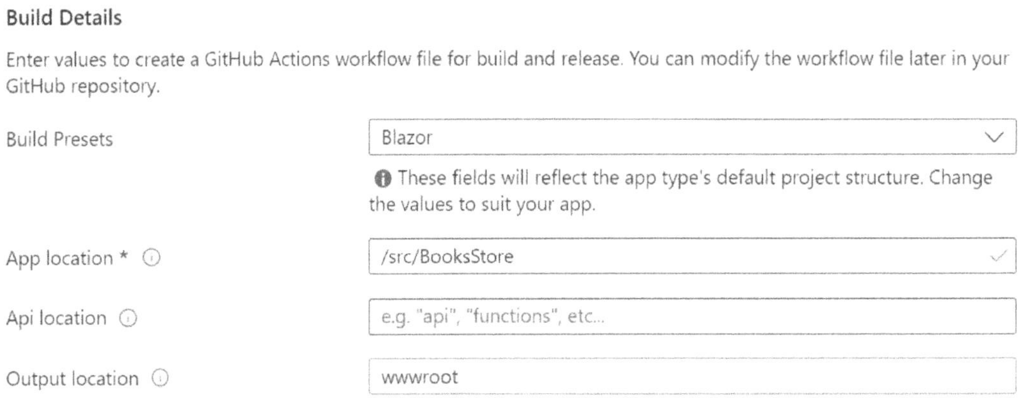

Build Details

Enter values to create a GitHub Actions workflow file for build and release. You can modify the workflow file later in your GitHub repository.

Build Presets | Blazor   ⌄ |

ⓘ These fields will reflect the app type's default project structure. Change
the values to suit your app.

App location * ⓘ | /src/BooksStore   ✓ |

Api location ⓘ | e.g. "api", "functions", etc... |

Output location ⓘ | wwwroot |

Figure 14.13 – Build Details of the Static Web App resource

10. Click on **Review + Create**, then **Create** again after the validation is finished.

11. Once the deployment is complete, click **Go to resource** so you can find the URL of the provisioned Azure Static Web App, and that's the URL we will use to access our application:

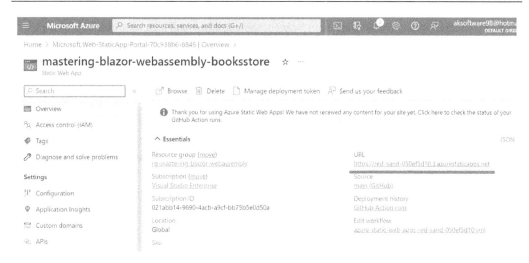

Figure 14.14 – The URL of the Azure Static Web App

That's it! The Azure Static Web App has modified your GitHub repository and created a GitHub Action workflow. This workflow will be triggered whenever you commit or merge a change to the `main` branch or the branch you defined during the creation of the Azure Static Web App.

If you navigate to the GitHub repository for the `BooksStore` source code and open the **Actions** tab, you should see a workflow that has been created by the Azure Static Web App:

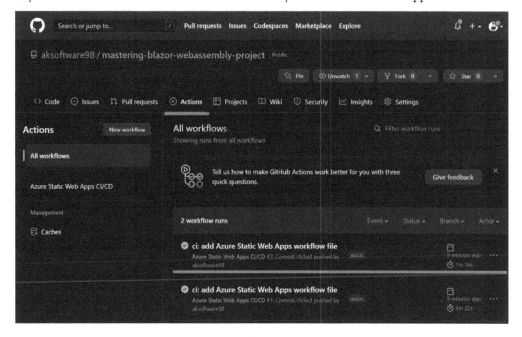

Figure 14.15 – GitHub Actions workflow created by the Azure Static Web App

With every change you submit to the `main` branch, the workflow will kick off to build and deploy the project automatically. In the *Blazor WebAssembly prerelease final checks* section, we talked about URL rewriting for Blazor WebAssembly standalone apps. So, the app should be working fine now, but if we try to navigate to `BASE_URL/counter`, for example, directly from the browser, we should get a **404 not found** page.

To resolve this, we need to instruct the Azure Static Web App to fall back any URL to `/index.html`. To do that, we need to add a new JSON file to the `BooksStore` project root directory called `staticwebapp.config.json` and, inside it, add the following JSON content:

```
{
  "navigationFallback": {
    "rewrite": "/index.html"
  }
}
```

This file contains the configuration that the Azure Static Web Apps service uses to handle requests. If you try to access `/book/form`, you will get a not found page because there is nothing on the server that corresponds to `/book/form`. Because for the Blazor WebAssembly SPA, the web file on the server is `index.html`, which is the single page of the app, other files are just the assets such as CSS, JS files, and so on. The `/book/form` URL is only valid within the Blazor app when it is running inside the browser. So, the host should automatically redirect you to `/index.html`, where the app will load and then navigate to the desired page on the client side.

If you commit this change now to the branch and navigate back to the **Actions** page of the GitHub repository, you will notice that the workflow has started to build and deploy the project again.

Wait until the workflow turns green and the status indicates success, which means if we navigate to the URL we obtained in *step 11*, we should see this fancy application shining on the screen:

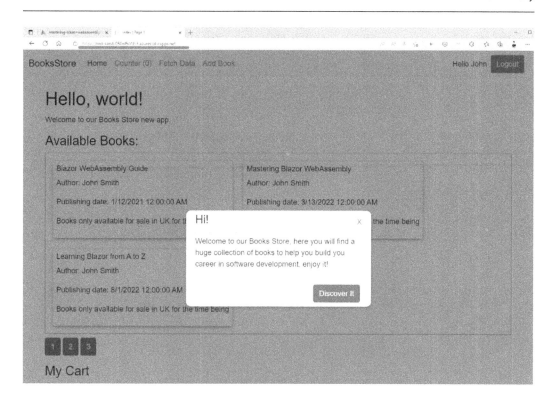

Figure 14.16 – The final app up and running on Azure Static Web Apps

Congratulations! You have built a full single-page application with Blazor WebAssembly, and you have learned how to build it securely and efficiently, how to test it, and more.

Summary

As this chapter comes to an end, our learning journey with this book is done. In this chapter, we learned how Blazor WebAssembly prepares an app for publishing, the steps it takes during this process, and how to configure them if needed. Then, we introduced the Blazor WebAsssembly ASP.NET Core Hosted project type and template, and then we deployed an ASP.NET Core Hosted Blazor project to Azure App Service. Finally, we deployed the project that we worked on during this book to the Azure Static Web Apps service.

After completing this chapter, you should be able to do the following:

- Understand AOT compilation and when to configure it
- Control publishing tasks such as trimming and compression

- Differentiate between Blazor WebAssembly Hosted and standalone projects
- Publish a Blazor WebAssembly ASP.NET Core Hosted project to Azure App Service
- Publish a Blazor WebAssembly standalone project to Azure Static Web Apps

You have put in some amazing effort and done great work. However, the journey of developing and exploring single-page applications in Blazor has just begun for you. In the next chapter, I will show you where to go from here.

Further reading

- *Host and deploy ASP.NET Core Blazor WebAssembly*: `https://learn. microsoft.com/en-us/aspnet/core/blazor/host-and-deploy/ webassembly?view=aspnetcore-7.0`

15
What's Next?

The journey of *Mastering Blazor WebAssembly* has come to an end. We have gone from explaining what Blazor WebAssembly is all the way to having an app running on Azure that is available for public access.

We have covered all the main topics that you may need in your day-to-day work with Blazor, but that is not everything; Blazor still has a lot to offer and explore.

This chapter will begin with a review of the additional components and functionalities that we have added to the book's project to help you consolidate your learning and use it to create more components and features for real-world applications. After that, we will briefly explore some other possibilities with Blazor and the next steps you can take from here.

This chapter will cover the following topics:

- Discovering more components and features in the book's project
- Using Blazor WebAssembly for mobile and desktop development
- Building Real-time applications with Blazor and SignalR
- Third-party UI components and packages
- Building your own apps

Technical requirements

In this chapter, we will cover features of the Blazor WebAssembly ASP.NET Core Hosted project to showcase more components and features in the `BooksStore` project. The completed version can be found in the GitHub repository of the book via the following URLs:

- The Blazor WebAssembly ASP.NET Core Hosted project: `https://github.com/PacktPublishing/Mastering-Blazor-WebAssembly/tree/main/Complete/AspNetCoreHosted/BooksStore`
- The Blazor WebAssembly standalone project: `https://github.com/PacktPublishing/Mastering-Blazor-WebAssembly/tree/main/Complete/Standalone`

Discovering more components and features in the book's project

This book has covered basic and advanced topics of Blazor using the practical approach of developing features and components for the `BooksStore` project, which simulates a real-world scenario. In this section, I will present more components and features that have been added to the project's repository, which will help you review what you have learned in this book and provide you with a useful resource that you can refer to when you encounter similar scenarios in your daily work.

The three features I will cover in this section are registering a user, rating and reviewing a book, and finally, how the admin can upload a cover file. Let's get started.

User registration

In *Chapter 9, Authenticating and Authorizing Users in Blazor*, in the *Building JWT custom authentication flow* section, we developed the authentication infrastructure and created a login page, but the users need to register too.

If you run the project and navigate to `/authentication/register`, the page will look as follows:

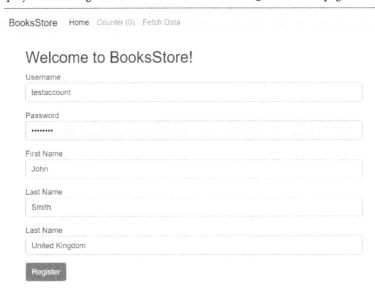

Figure 15.1 – User registration page

Here are the overall steps for building the register feature:

1. Build the register model that will carry the registration details of the user. The model named `RegisterUserRequest`, in the ASP.NET Core Hosted project, is in the `BooksStore.`

Shared project, but for the standalone project, the model can be found in the Models folder of the BooksStore project.

2. Write the method in AuthenticationService that will call the /authentication/register endpoint in the API to submit the registration details. In the ASP.NET Core Hosted and standalone projects, the logic can be found in IAuthenticationService.cs and AuthenticationService.cs in the Blazor WebAssembly project.

3. Create a page called Register.razor in the Pages/Authentication folder in the Blazor WebAssembly project. This page contains the user registration form and the full logic of submitting the data to the API.

Going over the steps and the files will help you practice a flow that's common for any operation in Blazor WebAssembly apps, and our project now supports a user registration operation.

Rating and reviewing

When we select a book from the home page, we will be taken to a book details page that shows more information and lets the user add the book to their cart, buy it, and more. It would be helpful if the buyers could give a book a rating from one to five stars, and write a comment so other buyers can know how good this book is.

In the BooksStore repository, I have added a Rating component and a BookRatingForm component to the BookDetails page, and the page now looks as follows:

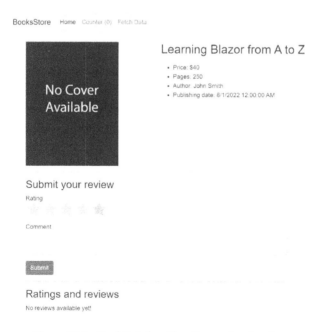

Figure 15.2 – BookDetails with rating and review form

That looks cool. The question is, how we can build this cute stars component and let the user submit a rating by clicking on any of the five stars?

The stars rating component can be found in the Shared folder of the Blazor WebAssembly project in the Rating.razor file.

I will explain briefly the logic of the Rating component, but first, we need to import the *Font Awesome* web font. Font Awesome provides scalable vector icons. We can use the icons and customize them using CSS. To bring Font Awesome into our project, we need to reference their CSS stylesheet in the index.html file of the Blazor project. You can download the CSS file or get an online reference via the following link: https://fontawesome.com/start.

The component has a for loop that renders five <i> elements with the fa and fa-star classes from Font Awesome, so the output will be five stars. There are multiple classes in the styles of the components: the CSS star class defines the size, the default color, and cursor pointer of the star; the CSS selected class sets the color of the star to gold; and the read-only class sets the cursor to default so the star doesn't indicate that it's clickable.

The component takes three parameters: StarCount of the int type, StarCountChanged of the EventCallback<int> type, and IsReadOnly of the bool type.

Each <i> element has the @onclick event handler. When a click event occurs, we check whether IsReadOnly is set to true. If it is, then we just return from the method, but if it's set to false, then we set StarCount to the number of the star that has been clicked and call the StarCountChanged EventCallback so the parent components can get the newly selected value from inside the Rating component. Each star builds a custom CSS classes list, so if the StarCount value is bigger than or equal to the star value, then we set the selected class to that star, and if IsReadOnly is set to true, then we append the read-only class to each star.

The logic is simple and very beneficial as a rating component is needed in many real-world cases.

Finally, to see how this component is being used in the full project, check out the BookRatingForm component in the Shared folder of the Blazor project, and the BookDetails page in the Pages/UserPage folder.

Uploading cover images

The last feature we will go over is uploading a cover file for the book. This feature is important because it allows you to upload files in Blazor WebAssembly.

The upload file logic can be found in the Blazor project in two places: the API call logic is in IBooksService and BooksHttpClientService is in the Services folder, and further logic can be found in the UploadBookCover.razor page in the Pages folder.

Blazor WebAssembly provides the InputFile component out of the box. InputFile allows us to pick single or multiple files. Each returned file is of the IBrowserFile type. IBrowserFile contains all the metadata of the file, such as the name and size, and it also contains the OpenReadStream method, which we can use to open a Stream for the selected file.

Now, let's talk about how the upload book cover feature works in the BooksStore project.

The project has a new page called **Upload Book Cover** that is accessible only by **Admin** users via /book/{BOOK_ID}/upload-cover.

If you log in as an admin and navigate to any book details, an **Upload Cover** button will appear at the top of the page that will redirect you to the **Upload Book Cover** page:

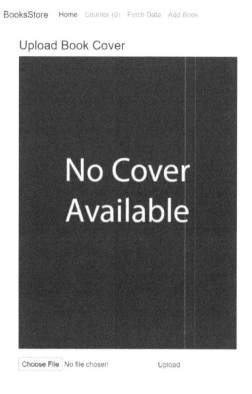

Figure 15.3 – Upload Book Cover page

The page simply retrieves the full book details and renders a cover (if one is available) and provides the user with an InputFile component and an **Upload** button.

The page has two methods. One is called OnChooseFile; this is assigned to the @onchange event of the InputFile component. It is triggered whenever the user chooses a file. OnChooseFile validates the extension of the file to make sure it's a valid image file, then assigns the chosen file to a variable of the IBrowserFile type.

The second method on the page is UploadCoverAsync; this method calls UploadBookCoverAsync inside the BooksHttpClientService method, which submits the file to the API.

The logic of submitting a file to an API is unlike the API calls we have done so far. Basically, the content of the HTTP request should be MultipartFormDataContent instead of JSON. This HTTP content allows uploading binary files inside the HTTP request body.

Inside BooksHttpClientService, there is the UploadBookCoverAsync method. This method receives a book ID as a parameter, a Stream object that represents the stream for the chosen file to be uploaded, and the filename:

```
public async Task UploadBookCoverAsync(string bookId, Stream stream,
string fileName)
{
    using var content = new MultipartFormDataContent();
    using var fileContent = new StreamContent(stream);
    fileContent.Headers.ContentDisposition = new
      ContentDispositionHeaderValue("form-data")
        {
            Name = "file",
            FileName = fileName
        };
    content.Add(fileContent);
    var response = await _httpClient.PostAsync
      ($"/books/{bookId}/cover", content);
    if (!response.IsSuccessStatusCode)
    {
        Var error = await response.Content
          .ReadFromJsonAsync<ApiErrorResponse>();
        Console.WriteLine(error);
    }
}
```

In the first line in the method, we create the content object of the MultipartFormDataContent type and an object of the StreamContent type called fileContent, which will contain the Stream of the file. After that, we use the headers of fileContent to set the name of the file in the form, which is file in this example.

Those were the three main features I wanted to share with you. In conclusion, going over these three components and features helps you retrieve the knowledge and put the skills we have developed throughout this book into practice again and shows you more capabilities and possibilities in Blazor WebAssembly to get you ready to start building your own apps.

With that, we can conclude the `BooksStore` project and all the Blazor WebAssembly topics. In the next sections, we will cover what else you can do with Blazor WebAssembly and provide some suggestions to get started with finding and implementing your first idea in Blazor WebAssembly.

Using Blazor WebAssembly for mobile and desktop development

The potential of Blazor WebAssembly is not limited only to the browser. With the power of .NET MAUI, you can develop apps that run on iOS, Android, Windows, and macOS and have access to the native capabilities of each device, such as storage, notifications, and camera.

By taking advantage of the rich Blazor UI capabilities and the components ecosystem, you can easily develop beautiful apps with astonishing UIs.

Check the Microsoft official documentation to get started: `https://learn.microsoft.com/en-us/aspnet/core/blazor/hybrid/tutorials/maui?view=aspnetcore-7.0&pivots=windows`.

Building real-time applications with Blazor and SignalR

SignalR is a real-time messaging framework that allows direct communication between the client and the server in both directions. With SignalR, the server can push content to the client instantly without the need for the client to request the data.

SignalR can be easily used in Blazor apps, and it enriches projects with many features, such as notifications, chat, and pushing updates from one browser to the other as they happen in real time. You can also develop live dashboards with SignalR as you send the statistics and charts updates from the server to the client.

To learn how to use SignalR in Blazor WebAssembly with the ASP.NET Core web API, check out the following URL: `https://learn.microsoft.com/en-us/aspnet/core/blazor/tutorials/signalr-blazor?view=aspnetcore-7.0&tabs=visual-studio&pivots=webassembly`.

Third-party UI components and packages

Throughout this book, we have learned about wonderful third-party packages built by the community, such as bUnit and Blazored.LocalStorage.

If you want to build advanced apps, developing your own components can be a bit costly. Luckily, there is a great set of wonderful UI component packages that provides you with a collection of components for all kinds of scenarios with a beautiful and customizable look.

Here is a list of my favorite components that I use from time to time:

- **MudBlazor**: This is definitely the most advanced and common UI package for Blazor. MudBlazor is an open source UI component library for Blazor applications. It provides a set of reusable and customizable components that are easy to integrate into your apps to build responsive and modern-looking software quickly. To learn more and get started with MudBlazor, check out their official website and documentation at `https://mudblazor.com`.

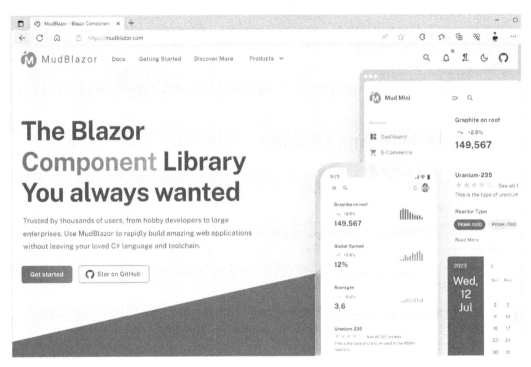

Figure 15.4 – MudBlazor official website

- **Blazorise**: This is another UI component library for Blazor that provides a variety of different components with a modern look. To learn more and get started, visit `https://blazorise.com/`.

- **Radzen Blazor Components**: This library contains a different set of components and is available at `https://blazor.radzen.com/`.

These three packages are fully open source and free to use, and they take the UI part away from you, so it's easy to create a new project and focus on a high-quality app with nice logic and features.

Building your own app

If you are reading this book and you are new to software development, then this is the right time to put all you have learned together in a real-world application. Go and discover the UI libraries mentioned in the previous section and create a new project.

Are you struggling to come up with an idea for an app? The best way to start is to work on a real-world app that you or someone around you will use. If you are not yet ready to develop apps to be used by others, develop apps for yourself. Here is a list of good ideas for you to get started with:

- Develop a time management app where you can track your work and study time

- Develop an expenses tracker app to track and manage all your expenses

- If any of your family or friends have a small business, consider developing an app to automate their work and track their accounting

- If you love racing, you can develop apps to extract insights about Formula 1 races using what you have learned and leverage an existing third-party API: `http://ergast.com/mrd/`

Those are just a set of suggestions, but of course, you can go with your own or get many more insights using tools such as ChatGPT or Bing.

Summary

In this, the final chapter of this book, we learned about creating some more features in the app we built. Those features will help you recap what we have learned and enable you to use the project we have built as a reference for your future work. You should by now know what else you can do with Blazor, such as mobile development with MAUI and real-time applications. In addition to some open source UI frameworks for Blazor that you can rely on while building your own apps, you also have some project ideas if you would like to do something cool on your own.

With these topics covered, you should have gained the following:

- A better understanding of the `BooksStore` project
- Knowledge of what else Blazor allows you to do, such as mobile development and more
- Knowledge of third-party packages and frameworks that accelerate your development
- Ideas to help you get started with building your first project

That's all! Congratulations on making it this far. I hope you enjoyed this journey as much as I did, and that you get the best out of it.

Index

Symbols

A

Packtpub.com

Subscribe to our online digital library for full access to over 7,000 books and videos, as well as industry leading tools to help you plan your personal development and advance your career. For more information, please visit our website.

Why subscribe?

- Spend less time learning and more time coding with practical eBooks and Videos from over 4,000 industry professionals

- Improve your learning with Skill Plans built especially for you

- Get a free eBook or video every month

- Fully searchable for easy access to vital information

- Copy and paste, print, and bookmark content

Did you know that Packt offers eBook versions of every book published, with PDF and ePub files available? You can upgrade to the eBook version at packtpub.com and as a print book customer, you are entitled to a discount on the eBook copy. Get in touch with us at customercare@packtpub.com for more details.

At www.packtpub.com, you can also read a collection of free technical articles, sign up for a range of free newsletters, and receive exclusive discounts and offers on Packt books and eBooks.

Other Books You May Enjoy

If you enjoyed this book, you may be interested in these other books by Packt:

Blazor WebAssembly By Example - Second Edition

Toi B. Wright

ISBN: 978-1-80324-185-2

- Discover the power of the C# language for both server-side and client-side web development
- Build your first Blazor WebAssembly application with the Blazor WebAssembly App project template
- Learn how to debug a Blazor WebAssembly app, and use ahead-of-time compilation before deploying it on Microsoft's cloud platform
- Use templated components and the Razor class library to build and share a modal dialog box
- Learn how to use JavaScript with Blazor WebAssembly
- Build a progressive web app (PWA) to enable native app-like performance and speed
- Secure a Blazor WebAssembly app using Azure Active Directory
- Gain experience with ASP.NET Web APIs by building a task manager app

Web Development with Blazor - Second Edition

Jimmy Engström

ISBN: 978-1-80324-149-4

- Understand the different technologies that can be used with Blazor, such as Blazor Server, Blazor WebAssembly, and Blazor Hybrid
- Find out how to build simple and advanced Blazor components
- Explore the differences between Blazor Server and Blazor WebAssembly projects
- Discover how Minimal APIs work and build your own API
- Explore existing JavaScript libraries in Blazor and JavaScript interoperability
- Learn techniques to debug your Blazor Server and Blazor WebAssembly applications
- Test Blazor components using bUnit

Packt is searching for authors like you

If you're interested in becoming an author for Packt, please visit `authors.packtpub.com` and apply today. We have worked with thousands of developers and tech professionals, just like you, to help them share their insight with the global tech community. You can make a general application, apply for a specific hot topic that we are recruiting an author for, or submit your own idea.

Share Your Thoughts

Now you've finished *Mastering Blazor WebAssembly*, we'd love to hear your thoughts! Scan the QR code below to go straight to the Amazon review page for this book and share your feedback or leave a review on the site that you purchased it from.

`https://packt.link/r/1-803-23510-1`

Your review is important to us and the tech community and will help us make sure we're delivering excellent quality content.

Download a free PDF copy of this book

Thanks for purchasing this book!

Do you like to read on the go but are unable to carry your print books everywhere?

Is your eBook purchase not compatible with the device of your choice?

Don't worry, now with every Packt book you get a DRM-free PDF version of that book at no cost.

Read anywhere, any place, on any device. Search, copy, and paste code from your favorite technical books directly into your application.

The perks don't stop there, you can get exclusive access to discounts, newsletters, and great free content in your inbox daily

Follow these simple steps to get the benefits:

1. Scan the QR code or visit the link below

https://packt.link/free-ebook/9781803235103

2. Submit your proof of purchase
3. That's it! We'll send your free PDF and other benefits to your email directly